Smart Inspection Systems

Techniques and Applications of Intelligent Vision

Full catalogue information on all books, journals and electronic products can be found on the Elsevier Science homepage at: http://www.elsevier.com

ELSEVIER PUBLICATIONS OF RELATED INTEREST

JOURNALS:

Advanced Engineering Informatics
Automatica
Computers in Industry
Control Engineering Practice
Engineering Applications of AI
Image and Vision Computing
International Journal of Machine Tool and Manufacture
Robotics and Autonomous Systems
Robotics and Computer-Integrated Manufacturing

Smart Inspection Systems
Techniques and Applications of Intelligent Vision

D.T. Pham
Manufacturing Engineering Centre
Cardiff University
Wales, UK

R.J. Alcock
Metropolis Informatics S.A.
Thessaloniki
Greece

ACADEMIC PRESS

An imprint of Elsevier Science
Amsterdam • Boston • London • New York • Oxford • Paris
San Diego • San Francisco • Singapore • Sydney • Tokyo

Academic Press
An Imprint of Elsevier Science
84 Theobald's Road, London WC1X 8RR, UK
http://www.academicpress.com

Academic Press
An Imprint of Elsevier Science
525 B Street, Suite 1900, San Diego, California 92101-4495, USA
http://www.academicpress.com

ISBN 0–12–554157–0

A catalogue record for this book is available from the British Library

Printed and bound in Great Britain by Biddles Ltd, *www.biddles.co.uk*

03 04 05 06 07   BL 9 8 7 6 5 4 3 2 1

Preface

Automated Visual Inspection (AVI) is a mechanised form of quality control normally achieved using one or more cameras connected to a computer system. Inspection is carried out to prevent unsatisfactory products from reaching the customer, particularly in situations where failed products can cause injury or even endanger life. Many humans are engaged in inspection tasks but, due to factors such as tiredness and boredom, their performance is often unreliable. In some cases, human inspection is not even possible when the part to be inspected is small, the production rates are high or there are hazards associated with inspecting the product. Therefore, AVI is gaining increasing importance in industry. Simultaneously, a growing amount of research has been aimed at incorporating artificial intelligence techniques into AVI systems to increase their capability.

The aim of this book is to enable engineers to understand the stages of AVI and how artificial intelligence can be employed in each one to create "smart" inspection systems. First, with the aid of examples, the book explains the application of both conventional and artificial intelligence techniques in AVI. Second, it covers the whole AVI process, progressing from illumination, through image enhancement, segmentation and feature extraction, to classification. Third, it provides case studies of implemented AVI systems and reviews of commercially-available inspection systems.

The book comprises seven chapters. Chapter One overviews the areas of AVI and artificial intelligence. The introduction to AVI discusses the requirements of AVI systems as well as its financial benefits. Seven popular artificial intelligence techniques are explained, namely expert systems, fuzzy logic, inductive learning, neural networks, genetic algorithms, simulated annealing and Tabu search.

Chapter Two covers image acquisition and enhancement. The key factor in image acquisition is the lighting. Common lighting system designs and light sources are detailed. The image enhancement methods presented range from traditional smoothing methods to the latest developments in smart enhancement.

Image segmentation is described in Chapter Three. A number of common segmentation techniques are covered, from the established methods of thresholding and edge detection to advanced segmentation based on artificial intelligence.

Chapter Four gives methods for feature extraction and selection. Several features are discussed, including first- and second-order features as well as window and object features. Statistical and classifier-based methods for selecting the optimal feature set are also described.

Classification techniques are the subject of Chapter Five. Four popular types of classifier are explained: Bayes' theorem classifiers, rule-based systems, neural networks and fuzzy classifiers. Synergistic classification, which combines the benefits of several individual classifiers, is also covered.

Chapter Six describes three applications of smart vision that have been implemented. The applications specified are the inspection of car engine seals and wood boards as well as the classification of textured images.

Finally, Chapter Seven reviews commercially-available inspection systems. Features of state-of-the-art vision systems are advanced cameras, intuitive development environments, intelligent algorithms and high performance. Many commercial inspection systems employ artificial intelligence to make the systems more effective, flexible and simple.

Each chapter concludes with exercises designed to reinforce and extend the reader's knowledge of the subject covered. Some of the exercises, based on a demonstration version of the ProVision image processing tool from Siemens contained in the CD ROM supplied with the book, also give the reader an opportunity to experience the key techniques discussed here.

Much of the work in this book derives from the AVI and artificial intelligence work carried out in the authors' award-winning Manufacturing Engineering Centre over the past 12 years. Several present and former members of the Centre have participated in AVI projects. They include Dr. E. Bayro-Corrochano, Dr. B. Cetiner, Dr. P.R. Drake, Mr. N.R. Jennings, Dr. M.S. Packianather, Dr. B. Peat, Dr. S. Sagiroglu and Dr. M. Yang, who are thanked for their contributions. Dr. B. Peat and Mr. A.R. Rowlands are also thanked for proof-reading the final manuscript.

The authors would like to acknowledge the financial support received for the work from the Engineering and Physical Sciences Research Council, the Department of Trade and Industry, the Teaching Company Directorate, the European Commission (BRITE-EURAM Programme) and the European Regional Development Fund (Knowledge-Based Manufacturing Centre, Innovative Technologies for Effective Enterprise and SUPERMAN projects). The work was performed in collaboration with industrial companies, including Europressings Ltd. (UK), Federal Mogul (UK), Finnish Wood Research Ltd. (Finland), Palla Textilwerke GMBH (Germany) and Ocas NV (Belgium). In particular, the authors wish to thank Dr. S. Baguley, Mr. M. Chung, Mr. G. Hardern, Mr. B. Holliday, Mr. J. Mackrory, Mr. U. Seidl and their colleagues at Siemens for the assistance given to the Centre.

The authors would also like to acknowledge the partners who have worked with them over the years in collaborative research efforts in AVI. They include Prof. P. Estevez (University of Chile), Mr. J. Gibbons (Ventek Inc., Eugene, Oregon), Mr. T. Lappalainen (VTT Technical Research Centre, Finland) and Dr. R. Stojanovic (University of Patras, Greece).

Dr. Alcock would personally like to express appreciation for the support of Metropolis Informatics S.A. (Thessaloniki, Greece), Prof. Y. Manolopoulos (Aristotle University of Thessaloniki), Dr. A.B. Chan and Dr. K. Lavangnananda.

Finally, the authors wish to thank Mr. N. Pinfield, Mr. H. van Dorssen, Mrs. L. Canderton and their colleagues at Academic Press and Elsevier Science for their expert support in the production of this book.

D.T. Pham
R.J. Alcock

Contents

Chapter 1

Automated Visual Inspection and Artificial Intelligence

Inspection is carried out in most manufacturing industries to ensure that low quality or defective products are not passed to the consumer. In financial terms, inspection is necessary because consumers who purchase unsatisfactory products are less likely to make a repeat purchase. More importantly, in the aerospace, automotive and food industries, failed products can cause injury or even fatal accidents. Many humans are engaged in inspection tasks but due to factors such as tiredness and boredom, their performance is often less than satisfactory. In some cases, human inspection is not even possible when the part to be inspected is very small or the production rates are very high. Thus, automated inspection is required. Another area where automated inspection is highly desirable is in the inspection of dangerous materials. These include inflammable, explosive or radioactive substances.

Automated Visual Inspection (AVI) is the automation of the quality control of manufactured products, normally achieved using a camera connected to a computer. AVI is considered to be a branch of industrial *machine vision* [Batchelor and Whelan, 1997]. Machine vision requires the integration of many aspects, such as lighting, cameras, handling equipment, human-computer interfaces and working practices and is not simply a matter of designing image processing algorithms. Industrial machine vision contrasts with high-level *computer vision*, which covers more theoretical aspects of artificial vision, including mimicking human or animal visual capabilities. Figure 1.1 shows a breakdown of artificial vision.

In modern manufacturing, quality is so important that AVI systems and human inspectors may be used together synergistically to achieve improved quality control [Sylla, 1993]. The machine vision system is used to inspect a large number of products rapidly. The human inspector can then perform slower but more detailed inspection on objects that the machine vision system considers to be borderline cases.

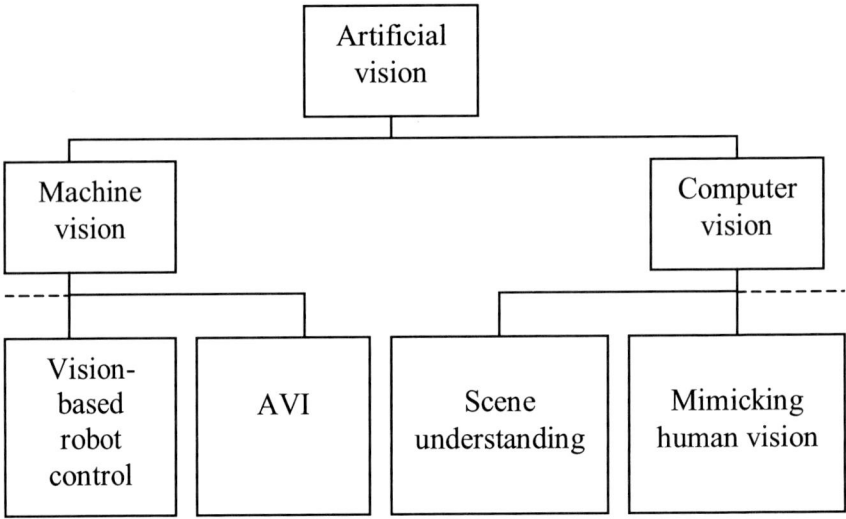

Figure 1.1 Examples of machine and computer vision

In AVI, many conventional image-processing functions, such as thresholding, edge detection and morphology, have been employed. However, much recent work has focussed on incorporating techniques from the area of artificial intelligence into the process. This book describes common artificial intelligence techniques and how they have been used in AVI.

This chapter gives the typical stages of the AVI process, explains common artificial intelligence techniques and outlines areas where artificial intelligence has been incorporated into AVI.

1.1 Automated Visual Inspection

AVI operates by employing a camera to acquire an image of the object being inspected and then utilising appropriate image processing hardware and software routines to find and classify areas of interest in the image. Figure 1.2 shows the set-up of an AVI system based around a central computer. In this system, the computer controls the camera, lighting and handling system. It also takes images acquired by the camera, analyses them using image processing routines and then issues an appropriate action to be performed by the handling system. Images from the inspected objects and the number of parts accepted and rejected may be displayed on a monitor or Visual Display Unit (VDU).

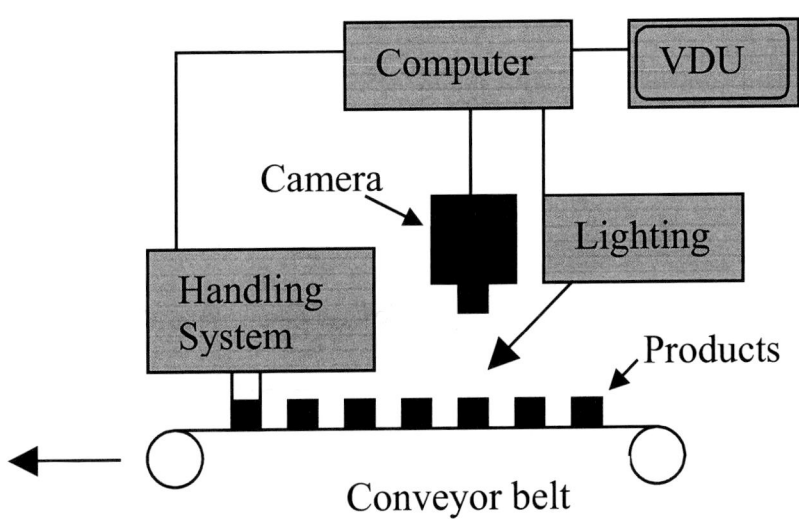

Figure 1.2 Typical AVI system structure

Generally, AVI involves the following processing stages (Figure 1.3):

- **Image acquisition** to obtain an image of the object to be inspected;

- **Image enhancement** to improve the quality of the acquired image, which facilitates later processing;

- **Segmentation** to divide the image into areas of interest and background. The result of this stage is called the segmented image, where *objects* represent the areas of interest;

- **Feature extraction** to calculate the values of parameters that describe each object;

- **Classification** to determine what is represented by each object.

Based on the classification, the parts are passed or failed. Accepted parts may then be graded. Another possible use of the classification information is as feedback for the production process. For example, it may be noticed that a particular type of defect is occurring frequently. This indicates that one of the machines or processes may not be operating optimally.

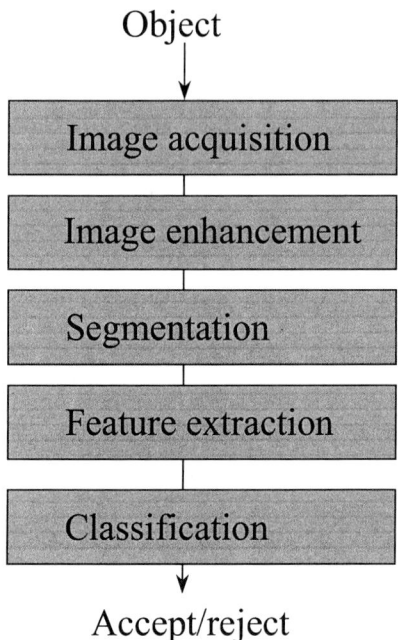

Figure 1.3 General AVI process structure

1.1.1 Practical Considerations for AVI Systems

Before implementing an AVI system, it is useful to consider some of the practical advice of Batchelor and Whelan [1997]:

- **System concept**. AVI systems should not be developed just for the sake of doing so. If a cheaper or easier alternative solution to the problem is available, then this is preferable. Also, neither humans nor vision systems will achieve 100% accuracy consistently. Thus, considering the importance of quality, the best solution may be to use humans and machines together.

- **Requirements**. It is important to start AVI system development with a clear system specification, detailing the customer requirements.

- **Design**. The system should be designed to be as simple as possible. This is called the principle of Occam's razor. If two systems are developed with the same performance, the one with the simplest implementation should be chosen. The justification for this is that a simple system has fewer components that can fail.

- **Implementation**. A larger part of the cost of the system development is spent on making the system work in a factory environment than on developing the image processing routines.

- **Visualisation**. It is beneficial to attach a VDU to the inspection system to inform personnel of its operation. Users and managers feel more confident with a system when they can visualise what it is doing.

The general requirements of an AVI system are that it should be:

- **Accurate**. It should improve upon human capabilities. The performance of human inspectors is usually well under 100%.

- **Fast**. The system should be able to work in real time. This is very important because objects in production lines arrive for inspection in rapid succession.

- **Robust**. The system should be insensitive to industrial environments. Such factors as variable light, vibrations and dust must be taken into consideration.

- **Complete**. The system should be able to identify, as well as locate, defects. The grade of an object can depend not only on the number and size of defects but also their type and location. In addition, the system should accumulate and make available statistical information about its operation for performance analysis purposes. If a defect, which is caused during production, occurs frequently then the system should notify the workers in the factory to correct the problem.

- **Flexible**. Ideally, the system should have some user-configurable modules so that it can be easily transferred from one product or production line to another. However, access to user-configurable parts of the system should be rigorously controlled and monitored.

- **Reliable**. If failure of the system is detected, a backup system or alarm will be required.

- **Maintainable.** The inspection equipment should be arranged so that all parts may easily be accessed. Also, computer programs should be written so that they are readable and easy to understand. Program code should contain a large number of comments and be logically structured.

- **Cost effective.** The cost of developing and running the system should be more than compensated for by its economic benefits. Often, the major cost in the development of an inspection system is not the hardware but the cost of employing developers to write dedicated software. However, considering the importance of quality in today's marketplace for acquiring and maintaining customers, the payback time for AVI systems can be short.

1.1.2 Financial Justifications for Automated Inspection

The UK Industrial Vision Association has produced a list of twenty-one financial justifications for using machine vision, which is available upon request [UKIVA, 2002]. Cost benefits can be divided into three broad categories:

- **Cost of quality.** Suppliers' quality can be monitored as well as the quality of finished products. Increased quality control should reduce the total cost of repair work required during the product guarantee.

- **Cost of materials.** The production of scrap materials will be reduced. Also, parts can be taken out of the production process as soon as they are found to be faulty so that good quality material further down the production line is not added to the faulty part.

- **Cost of labour.** This is generally a negligible effect and not a key motivator in the adoption of AVI.

One of the problems in the widespread acceptance of automated inspection systems is that companies who have installed them are unwilling to release details of their system or the cost savings because they want to maintain a competitive advantage over their rivals. For this reason, the European Union funded a programme to finance machine vision installations provided that the results of each project could be made public [Soini, 2000; HPCN-TTN, 2002]. Several machine vision projects have already yielded promising results. In the field of inspection, a bottle sorting system has been installed in a Finnish brewery. The system has increased line capacity from 4000 to 7000 crates per shift and gives cost savings equivalent to €50,000 per month.

1.2 Applications of AVI

Automated visual inspection has been applied to a wide range of products [Newman and Jain, 1995]. Due to the long set-up time for inspection systems, AVI is suited to tasks where a large number of products of the same type are made in a production-line environment. Table 1.1 gives examples of items for which automated inspection has been employed.

Demant et al. [1999] detailed three main applications areas of AVI systems in their practical guide to automated inspection:

- **Mark identification.** Marks on products that need to be checked include bar codes and printed characters on labels. For the checking of characters, special Optical Character Recognition (OCR) algorithms are required to determine what words have been printed.

- **Dimension checking.** Product dimensions can be verified as well as distances between sub-components on a part. Then, it can be determined if the part is manufactured to specification.

- **Presence verification.** On an assembly, it is normally required to check if all parts are present and correctly positioned.

Whilst it is relatively simple for humans to look at an object and determine its quality, it is a complex process to write image processing algorithms to perform the same task. Inspection systems work well when the product they are dealing with does not have a complex shape and is relatively uniform from one product to the next. Therefore, AVI is well suited to the inspection of objects such as car engine seals. Natural products, such as wood and poultry, can differ significantly from one to the next. Thus, whilst it is possible to inspect natural products, more complex inspection set-ups are required.

The most mature of AVI tasks is that of inspecting printed circuit boards (PCBs) [Moganti et al., 1996]. There are several reasons for the popularity of AVI in this area. First, PCBs are man made and so are regular. Second, their rate of production is very high, making their inspection virtually impossible for humans. Third, the quality requirements in the manufacture of PCBs are very high. Yu et al. [1988] found that the accuracy of human inspectors on multi-layered boards did not exceed 70%.

Object	Authors	Year
Apples	Wen and Tao	1999
Automobile axles	Romanchik	2001
Automotive compressor parts	Kang et al.	1999
Carpet	Wang et al.	1997
Castings	Tsai and Tseng	1999
Catfish	Korel et al.	2001
Ceramic dishes	Vivas et al.	1999
Corn kernels	Ni et al.	1997
Cotton	Tantaswadi et al.	1999
Electric plates	Lahajnar et al.	2002
Fish	Hu et al.	1998
Glass bottles	Hamad et al.	1998
Grain	Majumdar and Jayas	1999
Knitted fabrics	Bradshaw	1995
Lace	Yazdi and King	1998
Leather	Kwak et al.	2000
LEDs	Fadzil and Weng	1998
Light bulbs	Thomas and Rodd	1994
Metal tubes	Truchetet et al.	1997
Mushrooms	Heinemann et al.	1994
Olives	Diaz et al.	2000
Pistachio nuts	Pearson and Toyofuku	2000
Pizza topping	Sun	2000
Potatoes	Zhou et al.	1998
Poultry	Chao et al.	2002
Printed circuit boards	Chen et al.	2001
Pulp	Duarte et al.	1999
Seeds	Urena et al.	2001
Semiconductors	Kameyama	1998
Solder joints	Kim et al.	1999
Steel	Wiltschi et al.	2000
Textiles	Tolba and Abu-Rezeq	1997
Tiles	Melvyn and Richard	2000
Web materials	Hajimowlana et al.	1999
Weld	Suga and Ishii	1998
Wood	Pham and Alcock	1998

Table 1.1 Examples of automated inspection applications

Inspection methods for PCBs can be divided into contact and non-contact methods. Contact methods include electrical testing that find short circuits and open connections. AVI is a non-contact form of inspection. Due to the high quality requirements of PCB inspection, it is recommended that contact and non-contact testing methods be combined [Moganti et al., 1996].

Automated visual inspection tasks can be divided into four types, relating to the complexity of the images employed for the task: *binary*, *grey-scale*, *colour* and *range* [Newman and Jain, 1995].

When inspection tasks use binary images, the object generates a silhouette in front of a black or white background. Figure 1.4 shows an example of a binary image. For images acquired by a grey-scale camera, a simple fixed threshold is used to create the binary image. An advantage of using binary images is that it simplifies the requirements for image acquisition and lighting. In the simplest case, the image can be employed to determine the presence or absence of the part. However, typically, the image will be used to analyse the size, shape or position of the part. Then, it can be determined whether the part deviates significantly from its specifications.

Figure 1.4 Example of a binary image

The use of grey-scale, or intensity, images widens the number of applications for automated inspection. In particular, it allows the inspection of surfaces for defects and the analysis of texture information. Figure 1.5 shows an example of a grey-scale image.

Figure 1.5 Example of a grey-scale image

Colour images are used in AVI when grey-scale images cannot give a suitable level of accuracy.   Areas where colour adds important extra information include packaging, clothing and food inspection. Computers are more effective than humans at distinguishing between colours and "remembering" exactly what a colour looks like.

In the past, grey-scale images have been utilised more than colour ones for industrial AVI for two reasons. First, there is less information to process. Second, grey-scale systems have been cheaper to purchase. However, technological advances have meant that grey-scale systems are increasingly being replaced by colour ones.

Inspection systems that use range information are useful for inspecting objects where 3D information is important. One method of 3D inspection is through the use of co-ordinate measuring machines (CMMs). However, CMMs are too slow for on-line inspection.   Obtaining 3D information from 2D images is a complex computational problem.

1.3 Artificial Intelligence

Artificial intelligence (AI) involves the development of computer programs that mimic some form of natural intelligence. Some of the most common AI techniques with industrial applications are *expert systems*, *fuzzy logic*, *inductive learning*, *neural networks*, *genetic algorithms*, *simulated annealing* and *Tabu search* [Pham et al., 1998]. These tools have been in existence for many years and have found numerous industrial uses including classification, control, data mining, design, diagnosis, modelling, optimisation and prediction.

1.3.1 Expert Systems

Expert systems are computer programs embodying knowledge about a narrow domain for solving problems related to that domain [Pham and Pham, 1988]. An expert system usually comprises two main elements, a *knowledge base* and an *inference mechanism*. In many cases, the knowledge base contains several "IF - THEN" rules but may also contain factual statements, frames, objects, procedures and cases. Expert systems based on rules are also called *rule-based systems*. Figure 1.6 shows the typical structure of an expert system. First, domain knowledge is acquired from an expert. Second, facts and rules from the expert are stored in the knowledge base. Third, during execution of the system, the inference engine manipulates the facts and rules. Finally, the results of the inference process are presented to the operator in a user-friendly, often graphical, format.

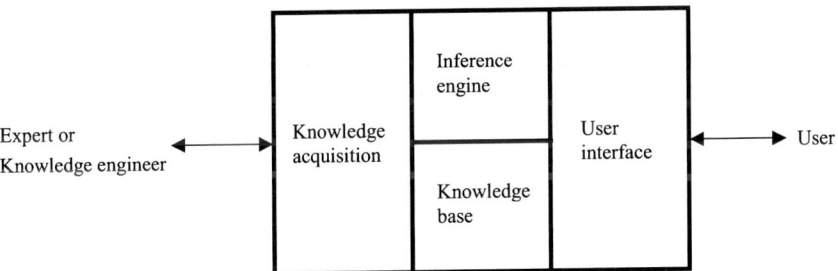

Figure 1.6 Typical structure of an expert system

An example of a rule that might be employed in inspection is:

if SIZE OF OBJECT > 25mm then REJECT

The inference mechanism manipulates the stored knowledge to produce solutions. This manipulation is performed according to a control procedure and search strategy. The control procedure may be either forward chaining or backward chaining, while the search strategy may be depth first, breadth first or best first.

Most expert systems are nowadays developed using programs known as "shells". These are ready-made expert systems with inference and knowledge-storage facilities but without the domain knowledge. Some sophisticated expert systems are constructed with the help of "development environments". The latter are more flexible than shells in that they also provide means for users to implement their own knowledge representation and inference methods.

Expert systems are now a very mature technology in artificial intelligence, with many commercial shells and development tools available to facilitate their construction. Consequently, once the domain knowledge to be incorporated in an expert system has been extracted, the process of building the system is relatively simple. The main problem in the development of expert systems is knowledge acquisition or the generation of the rules in the knowledge base.

1.3.2 Fuzzy Logic

A disadvantage of ordinary rule-based expert systems is that they cannot handle new situations not covered explicitly in their knowledge bases. These rule-based systems are unable to produce conclusions when such situations are encountered. Therefore, they do not exhibit a gradual reduction in performance when faced with unfamiliar problems, as human experts would.

The use of fuzzy logic, which reflects the qualitative and inexact nature of human reasoning, can enable expert systems to be more resilient [Nguyen and Walker, 1999]. With fuzzy logic, the precise value of a variable is replaced by a linguistic description, the meaning of which is represented by a fuzzy set, and inferencing is carried out based on this representation. Knowledge in an expert system employing fuzzy logic can be expressed as qualitative statements or fuzzy rules. For instance, in the rule given previously, the value of a fuzzy descriptor, such as the word large, might replace the value ">25". Thus, the rule would become:

if SIZE OF OBJECT IS LARGE then REJECT

A reasoning procedure, known as the compositional rule of inference, enables conclusions to be drawn by generalisation (extrapolation or interpolation) from the qualitative information stored in the knowledge base. In rule-based expert systems, the compositional rule of inference is the equivalent of the modus-ponens rule:

if A then B

A is TRUE => B is TRUE

The key aspect of fuzzy logic is the membership function. Figure 1.7 gives examples of membership functions to describe object size. Here, the fuzzy descriptors used to characterise the size are small, medium and large. It can be seen that if an object is 30mm long, it is definitely large. However, if an object is 25mm long, it can be described as both medium and large to a certain extent. The membership functions used in the example are triangular. It is also possible to employ trapezoidal, curved or other shapes of membership functions.

One of the main problems in fuzzy logic is to determine effective membership functions. One system developed for this is called WINROSA. WINROSA, which has been employed commercially, is based on the fuzzy ROSA (Rule Oriented Statistical Analysis) technique. Fuzzy ROSA was developed at the University of Dortmund in Germany [Krone and Teuber, 1996]. The generated rules can be integrated into many existing fuzzy shells, such as those in the commercial packages DataEngine, FuzzyTech and Matlab.

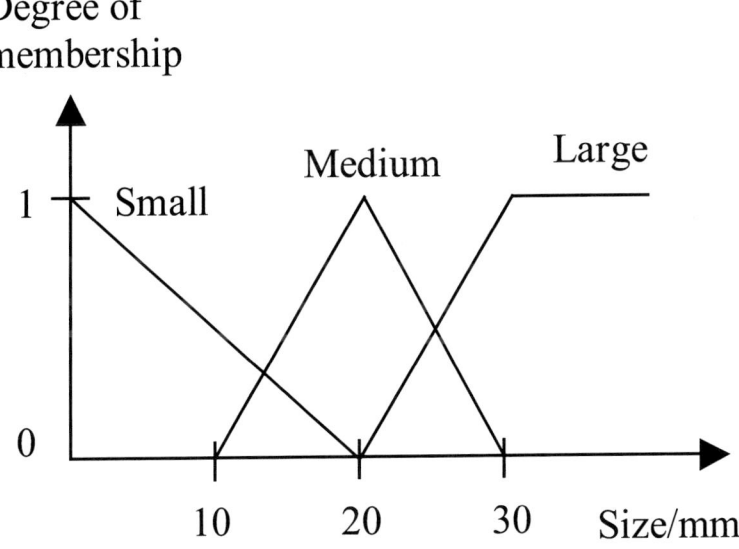

Figure 1.7 Example fuzzy membership functions for object size

1.3.3 Inductive Learning

The acquisition of domain knowledge for the knowledge base of an expert system is generally a major task. In many cases, it has proved a bottleneck in the process of constructing an expert system. Automatic knowledge acquisition techniques have been developed to address this problem. Inductive learning, in this case the extraction of knowledge in the form of IF-THEN rules (or an equivalent decision tree), is an automatic technique for knowledge acquisition. An inductive learning program usually requires a set of examples as input. Each example is characterised by the values of a number of attributes and the class to which it belongs.

In the *tree-based* approach to inductive learning, the program builds a decision tree that correctly classifies the training example set. Attributes are selected according to some strategy (for example, to maximise the information gain) to divide the original example set into subsets. The tree represents the knowledge generalised from the specific examples in the set. Figure 1.8 shows a simple example of how a tree may be used to classify citrus fruit. It should be noted that if size were used at the root of the tree instead of colour, a different classification performance might be expected.

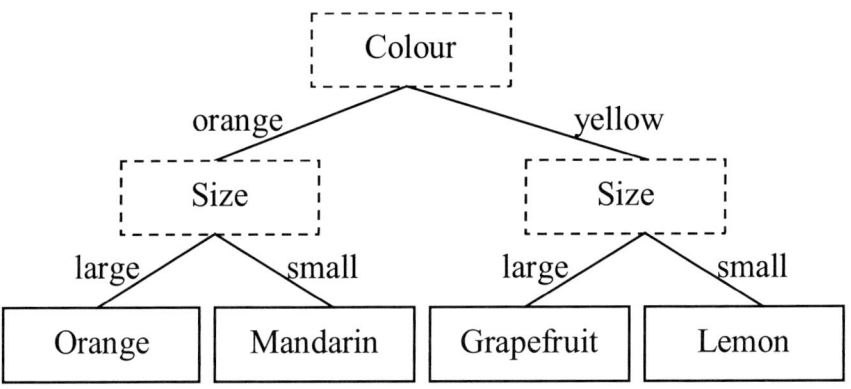

Figure 1.8 Decision tree for classifying citrus fruit

In the *rule-based* approach, the inductive learning program attempts to find groups of attributes uniquely shared by examples in given classes and forms rules with the IF part as combinations of those attributes and the THEN part as the classes. After a new rule is formed, the program removes correctly classified examples from consideration and stops when rules have been formed to classify all examples in the training set.

Several inductive learning algorithms and families of algorithms have been developed. These include ID3, AQ and RULES. The ID3 algorithm, developed by Quinlan [1983], produces a decision tree. At each node of the tree, an attribute is selected and examples are split according to the value that they have for that attribute. The attribute to employ for the split is based on its *entropy* value, as given in Equation (1.1). In later work, the ID3 algorithm was improved to become C4.5 [Quinlan, 1993].

$$\text{Entropy(attribute)} = -(P_+ \log_2 P_+) - (P_- \log_2 P_-) \qquad (1.1)$$

where P_+ is the proportion of examples that were correctly classified just using the specified attribute and P_- is the proportion incorrectly classified. For implementation, 0.log(0) is taken to be zero.

The AQ algorithm, created by Michalski et al. [1986], uses the rule-based approach to inductive learning. Combinations of attributes and their values create rules. The rules are searched, from general to specific cases. A rule is considered to be more general if it has fewer conditions. A generated rule that does not misclassify any examples in the training set is kept. Thus, the derived rules give a 100% performance on the training set. To deal with numerical attributes, the attribute values need to be quantised so that they are like nominal attribute values. AQ has been through a number of revisions. In 1999, AQ18 was released [Kaufman and Michalski, 1999].

An extension to AQ that was designed to handle noise in the data is the CN2 algorithm [Clark and Niblett, 1989]. CN2 keeps rules that classify some examples incorrectly. It does not obtain a 100% accuracy on the training data but can give a better performance on unseen data. The formula used to assess the quality of a rule is:

$$\frac{c+1}{c+i+n} \qquad (1.2)$$

where c is the number of examples classified correctly by the rule, i is the number incorrectly classified by the rule and n is the number of classes.

Pham and Aksoy [1993; 1995a, b] developed the first three algorithms in the RULES (RUle Extraction System) family of programs. These programs were called RULES-1, 2 and 3. Later, the rule forming procedure of RULES-3 was improved by Pham and Dimov [1997a] and the new algorithm was called RULES-3 PLUS. Rules are generated and those with the highest 'H measure' are kept. The H measure is calculated as:

$$H = \sqrt{\frac{E^c}{E}} \times [2 - 2\sqrt{\frac{E_i^c E_i}{E^c E}} - 2\sqrt{(1 - \frac{E_i^c}{E^c})(1 - \frac{E_i}{E})}]$$ (1.3)

where:

E is the total number of instances;

E^c is the total number of instances covered by the rule (whether correctly or incorrectly classified);

E_i^c is the number of instances covered by the rule and belonging to target class i (correctly classified);

E_i is the number of instances in the training set belonging to target class i.

The first incremental learning algorithm in the RULES family was RULES-4 [Pham and Dimov, 1997b]. Incremental learning is useful in cases where not all the training data is known at the beginning of the learning process. RULES-4 employs a Short Term Memory (STM) to store training examples when they become available. The STM has a user-specified size called the STM size. When the STM is full, the RULES-3 PLUS procedure is used to generate rules.

Alcock and Manolopoulos [1999] carried out experiments on RULES-4 to determine optimum values for the parameters. RULES-4 has three parameters that need to be set: the STM size, number of quantisation levels (Q) and the noise level (NL). It was seen that as the STM size increases, the performance improves. The most sensitive of the three parameters was Q. Choosing an appropriate value for Q has a large effect on classification accuracy.

1.3.4 Neural Networks

There are many classifier architectures which are covered by the term Artificial Neural Networks (ANNs) [Pham and Liu, 1999]. These networks are models of the brain in that they have a learning ability and a parallel distributed architecture.

Like inductive learning programs, neural networks can capture domain knowledge from examples. However, they do not archive the acquired knowledge in an explicit form such as rules or decision tress. Neural networks are considered a 'black box' solution since they can yield good answers to problems even though they cannot provide an explanation of why they gave a particular answer. They also have a good generalisation capability, as with fuzzy expert systems.

A neural network is a computational model of the brain. Neural network models usually assume that computation is distributed over several simple units called neurons that are interconnected and operate in parallel (hence, neural networks are also called parallel-distributed-processing or connectionist systems).

The most popular neural network is the Multi-Layer Perceptron (MLP) which is a feedforward network: all signals flow in a single direction from the input to the output of the network. Feedforward networks perform a static mapping between an input space and an output space. The output at a given instant is a function only of the input at that instant.

Recurrent networks, where the outputs of some neurons are fed back to the same neurons or to neurons in previous layers, are said to have a dynamic memory. The output of such networks at a given instant reflects the current input as well as previous inputs and outputs.

Neural networks "learn" a mapping by training. Some neural networks can be trained by being presented with typical input patterns and the corresponding target output patterns. The error between the actual and expected outputs is used to modify the strengths, or weights, of the connections between the neurons. This method of training is known as supervised training.

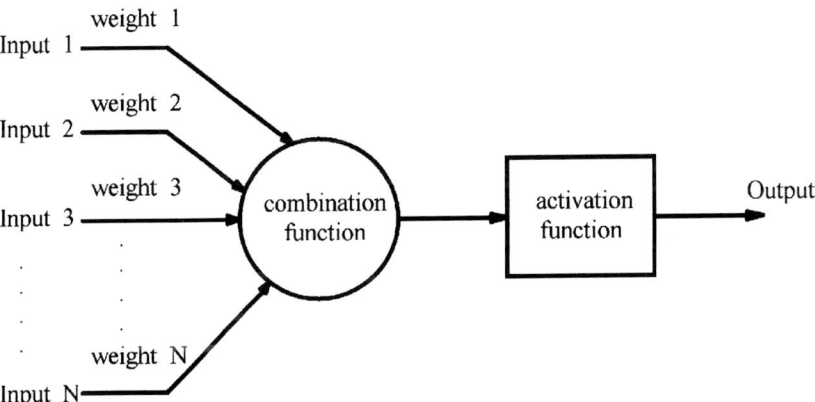

Figure 1.9 Artificial neuron

The MLP is claimed to be the ANN with the nearest architecture to that of the brain. It consists of a number of artificial neurons. Figure 1.9 displays a neuron in a MLP. The output of each neuron is determined by the sum of its inputs and an activation function. In the figure, the N inputs to the neuron are labelled from 1 to N. These are multiplied by weights 1 to N, respectively, and combined (normally by summation).   By passing the result of the combination through an activation function, the final output of the neuron is derived.

In the MLP, the neurons are linked together by weighted connections. An example of this is shown in Figure 1.10, where circles represent neurons and lines represent weighted connections. The example network has three layers: an input layer, an output layer and an intermediate or hidden layer. Neurons called bias neurons, that have a constant output of one, are also included. The determination of optimal values for the weights is the learning phase of the network. The learning algorithm most commonly utilised to determine these weights is back-propagation (BP). This algorithm propagates the error from the output neurons and computes the weight modifications for the neurons in the hidden layers.

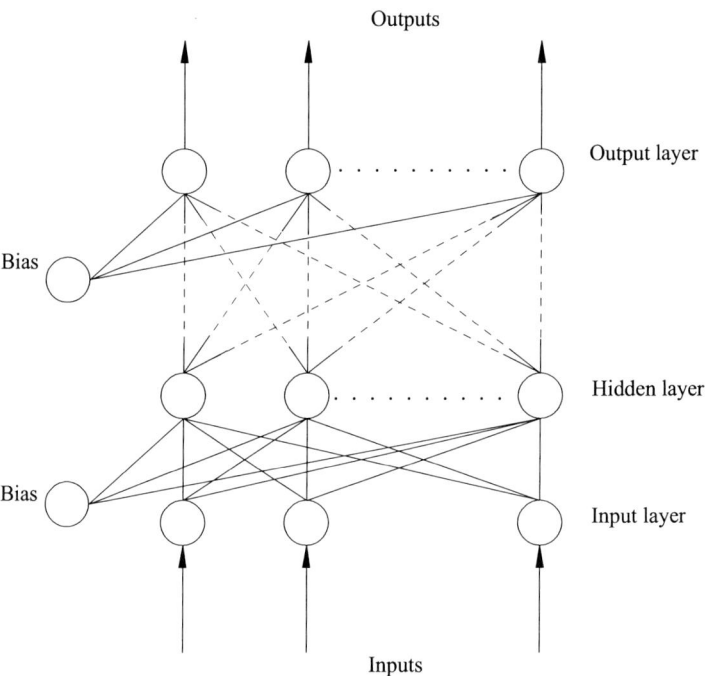

Figure 1.10 MLP neural network

Neurons in the input layer only act as buffers for distributing the input signals x_i to the neurons in the hidden layer. Each neuron j in the hidden layer sums its input signals x_i, after multiplying them with the strengths of the respective connections w_{ji} from the input layer, and computes its output y_j as a function f of the sum:

$$y_j = f(\Sigma w_{ji} x_i) \qquad (1.4)$$

Function f can be a simple threshold or a sigmoidal, hyperbolic tangent or radial basis function. The output of the neurons in the output layer is calculated similarly. A common activation function is the sigmoidal function:

$$f(x) = \frac{1}{1 + e^{-x}} \qquad (1.5)$$

The backpropagation algorithm gives the change Δw_{ji} in the weight of a connection between neurons i and j as:

$$\Delta w_{ji} = \eta \delta_j x_i \qquad (1.6)$$

where η is a parameter called the learning rate and δ_j is a factor depending upon whether neuron j is an output neuron or a hidden neuron. For output neurons:

$$\delta_j = \frac{df}{dnet_j}(y_j^{(t)} - y_j) \qquad (1.7)$$

and for hidden neurons:

$$\delta_j = \frac{df}{dnet_j} \sum_q w_{qj} \delta_q \qquad (1.8)$$

In equation (1.7), net_j is the total weighted sum of input signals to neuron j and $y_j^{(t)}$ is the target output for neuron j.

As there are no target outputs for hidden neurons, in equation (1.8), the difference between the target and actual output of a hidden neuron j is replaced by the weighted sum of the δ_q terms already obtained for neurons q connected to the output of j. Thus, iteratively, beginning with the output layer, the δ term is computed for all neurons in all layers and weight updates determined for all connections. The weight updating process can take place after the presentation of each training pattern (pattern-based training) or after the presentation of the whole set of training patterns (batch training). In either case, a training epoch is complete when all the training patterns have been presented once to the MLP.

For all but the most trivial problems, several epochs are required for the MLP to be properly trained. A commonly-adopted method to speed up the training is to add a "momentum" term to equation (1.6), which effectively lets the previous weight change influence the new weight change, viz.:

$$\Delta w_{ji}(k+1) = \eta \delta_j x_i + \alpha \Delta w_{ji}(k) \tag{1.9}$$

where $\Delta w_{ji}(k+1)$ and $\Delta w_{ji}(k)$ are weight changes in epochs $(k+1)$ and (k), respectively, and α is the "momentum" coefficient.

For interested readers, further details and code for the MLP using BP training can be found in [Pham and Liu, 1999; Eberhart and Dobbins, 1990].

Some neural networks are trained using unsupervised learning, where only the input patterns are provided during training. These networks learn automatically to cluster patterns into groups with similar features. Examples of such networks are the Kohonen Self-Organising Feature Map (SOFM) and Adaptive Resonance Theory (ART) [Pham and Chan, 1998].

A basic SOFM contains a two-dimensional map of neurons. Figure 1.11 gives an illustration of a 3x3 Kohonen SOFM. The neurons in a Kohonen map (and in an ART network) are different from those in a MLP network. MLP neurons process the inputs using mathematical functions. In the Kohonen map, the neurons simply store representative feature vectors. Each neuron has the same dimensionality as the input feature vector. During training, when patterns are presented, the neuron is found which is closest to the input pattern, normally calculated using Euclidean distance. The weights of that neuron are updated, as are its neighbours but to a decreasing extent according to their distance away from the selected neuron. Over a series of iterations, the weights of the neurons are adapted so that, finally, different areas of the map correspond to different types of input patterns. A variant of the SOFM network is a supervised network called the Learning Vector Quantisation (LVQ) network [Pham and Oztemel, 1996].

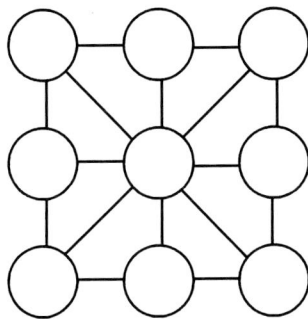

Figure 1.11 3x3 Kohonen SOFM

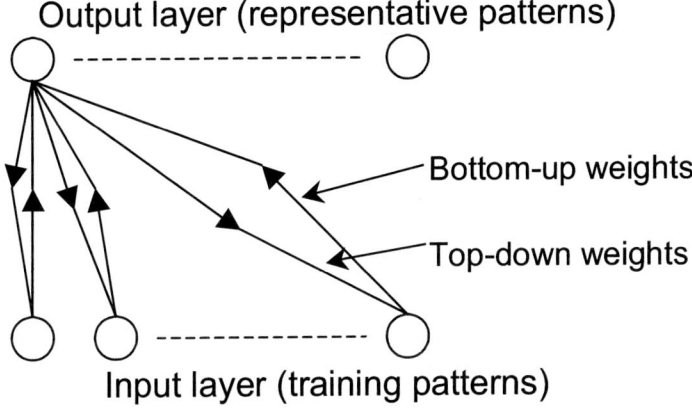

Output layer (representative patterns)

Bottom-up weights

Top-down weights

Input layer (training patterns)

Figure 1.12 ART-1 network

ART aims to overcome the "stability-plasticity dilemma". Normally, after a neural network is trained, it becomes static and is unable to learn from new input patterns. This contrasts with humans, who are able to update their knowledge continuously. In ART, the first input pattern presented creates a new neuron. Subsequently, patterns are compared with existing neurons. If they are found to be sufficiently close to an existing neuron (according to some similarity criterion) the pattern is assigned to that neuron and the neuron's weights are updated. Otherwise, a new neuron is generated for that pattern. Thus, it is not certain how many neurons will be generated during training. This is different from other kinds of neural networks, where the number of neurons is fixed at the beginning of training. An illustration of ART-1, an early type of ART network, is shown in Figure 1.12. ART-2 has a similar operating principle to ART-1 but a more complex architecture, enabling it to handle noisy analogue inputs.

Another popular development in neural networks is that of Radial Basis Functions (RBFs) [Schalkoff, 1997]. One particular advantage that RBFs have over MLPs is that they do not need to be retrained completely when new training data becomes available. A basic RBF network has three layers: input, hidden and output. The dimensionality of the hidden layer should be selected according to the problem. The connections between the input and hidden layers are non-linear but the hidden-to-output layer connections are linear. The idea of the hidden layer is to transform non-linearly separable patterns into linearly separable patterns so that the output layer can easily deal with them. The training phase for the RBF network involves finding the appropriate non-linear functions for the hidden layer. The output of hidden neuron i in a RBF network has the following form:

$$w_i \varphi(\|x - x_i\|) \qquad (1.10)$$

where w_i is the weight for hidden neuron i, ϕ is a non-linear function, x is the input pattern, x_i is the pattern stored in hidden neuron i and $\| \bullet \|$ is the distance function. The function ϕ is often chosen to be a Gaussian.

The problem of training a RBF network is thus one of determining optimal values for w_i and x_i. To find values for all x_i, the k-means clustering algorithm or the SOFM could be employed. One problem that must be avoided when developing a RBF network is to use too many hidden neurons. This could cause an effect called *overfitting*, which means that the training data is learnt too well and a poor generalisation performance is achieved.

1.3.5 Genetic Algorithms, Simulated Annealing and Tabu Search

Intelligent optimisation techniques, such as genetic algorithms, simulated annealing and Tabu search, look for the answer to a problem amongst a large number of possible solutions [Pham and Karaboga, 2000]. A classic example of an optimisation problem is the travelling salesman problem. In this problem, a salesman needs to visit a number of cities in the shortest possible distance. As all the cities are known, all possible route combinations are also known. Thus, the problem is one of searching through all the possible routes. When the number of cities becomes large, the number of route combinations increases excessively. Thus, a fast search technique is required to find the optimal route. Optimisation techniques are used to perform fast searching. Common AI optimisation techniques include genetic algorithms, simulated annealing and Tabu search.

Genetic algorithms (GAs) are currently one of the most popular optimisation techniques. GAs are computer programs that are based on the theory of natural evolution [Goldberg, 1989]. The basis of this theory is that animals are solutions to the problem of survival. Only animals that are able to survive can reproduce and thus only fit animals can pass their genes to the next generation. Over successive generations, the collective fitness of the overall population increases.

Genetic algorithms can be applied to problems where it is required to search for a solution. Possible solutions are represented as *genes* or *chromosomes*. A *fitness function* is used to determine which chromosomes survive to the next iteration. New chromosomes replace ones with a low fitness. The chromosomes in a GA are normally represented in binary form, that is, containing just ones and zeros. At any one time, a *population* of chromosomes is stored. Reproduction is performed by an operation called *crossover*, where two chromosomes are combined to create a new one. Diversity is maintained in the population by the *mutation* operation that generates a new chromosome by selecting an existing one and inverting one of its elements. The frequency the mutation operation is applied is determined by a user-specified parameter called the *mutation rate*. Figure 1.13 illustrates the operations of mutation and crossover.

010110   Original chromosome
↓
010010   Mutated chromosome

010101   Original chromosome 1
↓↓↓
110101   New chromosome after crossover
↑↑ ↑
110011   Original chromosome 2

Figure 1.13 Mutation and crossover

Figure 1.14 shows the typical operation of a genetic algorithm. First, an initial population of chromosomes is developed. Second, a cycle is entered where new populations are generated and evaluated. The cycle starts by evaluating all chromosomes in the population on the basis of their fitness values. Next, new chromosomes are generated, which will be evaluated at the start of the next cycle. The cycle is terminated after either a fixed number of iterations or a pre-specified fitness level is reached.

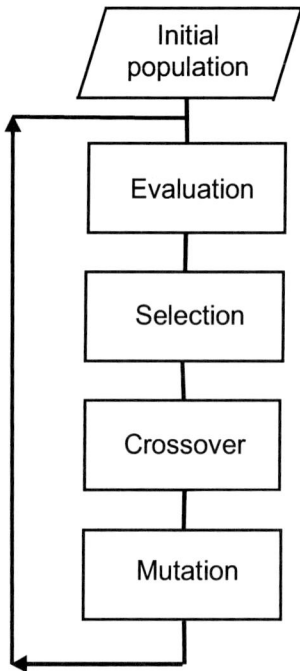

Figure 1.14 Typical operation of a genetic algorithm

GAs have found many applications. For example, Alcock and Manolopoulos [1999] showed that genetic algorithms could be employed for inductive learning. They developed a system called GAIL (Genetic Algorithm for Inductive Learning). GAIL searches through possible rules, assigning a fitness value to each one. During learning, the GA maintains a population of rules with the highest fitness level. New rules are generated using the crossover and mutation operations. In experiments, GAIL was found to give small rule sets with good accuracy. Lavangnananda [2001] developed synergistic GAIL.

Simulated annealing is an optimisation technique that is based on the process of metal annealing. In annealing, a metal is heated to a high temperature and then gradually cooled. In this way, the atoms in the metal settle down to form a solid structure that is stronger than the original metal. In simulated annealing, the temperature is a variable that is initially set to a high value and reduced over time. First, a random solution to the problem is generated. Then, based on this solution, a new one is created. If the new solution is better than the old one then it is accepted. If it is worse then the solution is accepted with a probability based on the temperature variable. The probability p of accepting a worse solution is calculated as:

$$p = e^{\frac{-\Delta E}{T}} \qquad (1.11)$$

where, ΔE is the change in the fitness of the solution and T is the current value of the temperature variable.

When the temperature variable is high, the solution changes rapidly. At lower temperatures, fewer changes are accepted. As the value of the temperature decreases, the solution becomes more stable and ideally will converge to the optimal value. One of the main differences between GAs and simulated annealing is that GAs concurrently handle a population of solutions, whereas simulated annealing deals with just one.

Tabu search was first developed in 1986 [Glover, 1986; Hansen, 1986]. Like GAs, Tabu search employs an evaluation function. At each iteration, the solution is chosen that gives the highest value for the evaluation function. A *Tabu list* is given to show which solutions can be reached from the current solution. Tabu search requires three strategies. The *forbidding strategy* determines what can enter the Tabu list whilst the *freeing strategy* determines what can leave. The *short-term strategy* controls the interaction of the first two strategies.

1.4 Artificial Intelligence in AVI

It can be argued that machine vision is by its nature intelligent because the interpretation of images requires intelligence. Humans find vision tasks trivial as the brain processes images subconsciously, without any need for deliberate thought. Conversely, interpretation of images is complex for computers but tasks depending on a large amount of known calculations, such as playing chess, are relatively simple. No computer system has so far been developed which can interpret all images. Rather, many machine vision systems have been built that deal with one specific object in a restricted environment.

Over recent years, a growing amount of research has been aimed at incorporating artificial intelligence techniques into AVI systems to increase their capability. In the following chapters of this book, several examples will be given where artificial intelligence has been employed for AVI. Table 1.2 shows the techniques for each stage of the AVI process that are described in this book.

	Conventional methods	Expert systems	Fuzzy logic	Inductive learning	Neural networks	Genetic algorithms, simulated annealing and Tabu search
Image acquisition	X	X				
Image enhancement	X		X		X	X
Segmentation	X		X		X	X
Feature selection	X		X	X	X	X
Classification	X	X	X	X	X	

Table 1.2 AVI stages and AI techniques

1.5 Summary

Automated visual inspection is becoming increasingly important for the inspection of many industrial products. It can be decomposed into the stages of image acquisition, image enhancement, segmentation, feature extraction and classification. Artificial intelligence techniques are being increasingly utilised. Common artificial intelligence techniques include expert systems, fuzzy logic, inductive learning, neural networks, genetic algorithms, simulated annealing and Tabu search. One or more of these methods has been used at each stage of the inspection process.

References

Alcock R.J. and Manolopoulos Y. (1999) Genetic Algorithms for Inductive Learning. In *Computation Intelligence and Applications*. (ed. Mastorakis N.) World Scientific, Singapore. pp. 85 - 90.

Batchelor B.G. and Whelan P.F. (1997) *Intelligent Vision Systems for Industry*. Springer-Verlag, Berlin and Heidelberg.

Bradshaw M. (1995) The Application of Machine Vision to the Automated Inspection of Knitted Fabrics. *Mechatronics*. Vol. 5, No. 2-3, pp. 233 - 243.

Chao K., Chen Y.R., Hruschka W.R. and Gwozdz F.B. (2002) On-line Inspection of Poultry Carcasses by a Dual-camera System. *Journal of Food Engineering*. Vol. 51, No. 3, pp. 185 - 192

Chen T.Q., Zhang J., Murphey Y.L., Zhou Y. (2001) A Smart Machine Vision System for PCB Inspection. *Lecture Notes in Computer Science*. No. 2070, pp. 513 - 518.

Clark P. and Niblett T. (1989) The CN2 Induction Algorithm. *Machine Learning*. Vol. 3, pp. 261 - 283.

Demant C., Streicher-Abel B. and Waszkewitz P. (1999) *Industrial Image Processing: Visual Quality Control in Manufacturing*. Springer-Verlag, Berlin.

Diaz R., Faus G., Blasco M., Blasco J. and Molto E. (2000) The Application of a Fast Algorithm for the Classification of Olives by Machine Vision. *Food Research International*. Vol. 33, No. 3, pp. 305 - 309.

Duarte F., Araujo H. and Dourado A. (1999) An Automatic System for Dirt in Pulp Inspection using Hierarchical Image Segmentation. *Computers and Industrial Engineering*. Vol. 37, No. 1-2, pp. 343 - 346.

Eberhart R.C. and Dobbins R.W. (1990) *Neural Networks PC Tools: A Practical Guide*. Academic Press, San Diego.

Fadzil M.H.A. and Weng C.J. (1998) LED Cosmetic Flaw Vision Inspection System. *Pattern Analysis and Applications*. Vol. 1, No. 1, pp. 62 - 70.

Glover F. (1986) Future Paths for Integer Programming and Links to Artificial Intelligence. *Computers and Operations Research*. Vol. 13, pp. 533 - 549.

Goldberg D.E. (1989) *Genetic Algorithms in Search, Optimization and Machine Learning.* Addison-Wesley, Reading, MA.

Hajimowlana S.H., Muscedere R., Jullien G.A. and Roberts J.W. (1999) An In-Camera Data Stream Processing System for Defect Detection in Web Inspection Tasks. *Real-time Imaging.* Vol. 5, No. 1, pp. 23 - 34.

Hamad D., Betrouni M., Biela P. and Postaire J.G. (1998) Neural Networks Inspection System for Glass Bottles Production: A Comparative Study. *Int. Journal of Pattern Recognition and Artificial Intelligence.* Vol. 12, No. 4, pp. 505 - 516.

Hansen P. (1986) The Steepest Ascent Mildest Descent Heuristic for Combinatorial Programming. *Conf. on Numerical Methods in Combinatorial Optimisation.* Capri, Italy.

Heinemann P.H., Hughes R., Morrow C.T., Sommer H.J., Beelman R.B. and West P.J. (1994) Grading of Mushrooms using a Machine Vision System. *Trans. of the ASAE.* Vol. 37, No. 5, pp. 1671 - 1677.

HPCN-TTN (2002) *High Performance Computing Network - Technology Transfer Nodes.* European Commission, Brussels. http://www.hpcn-ttn.org.

Hu B.G., Gosine R.G., Cao L.X. and De Silva C.W. (1998) Application of a Fuzzy Classifier in Computer Grading of Fish Products. *IEEE Trans. on Fuzzy Systems.* Vol. 6, No. 1, pp. 144 - 152.

Kameyama K., Kosugi Y., Okahashi T. and Izumita M. (1998) Automatic Defect Classification in Visual Inspection of Semiconductors using Neural Networks. *IEICE Trans. on Information and Systems.* Vol. E81D, No. 11, pp. 1261 - 1271.

Kang D.K., Chung Y.K., Doh W.R., Jung W. and Park S.B. (1999) Applying Object Modelling Technique to Automated Visual Inspection of Automotive Compressor Parts Omission. *Int. Journal of Machine Tools and Manufacture.* Vol. 39, No. 11, pp. 1779 - 1792.

Kaufman K.A. and Michalski R.S. (1999) Learning from Inconsistent and Noisy Data: The AQ18 Approach. *Proc. 11th Int. Symp. on Methodologies for Intelligent Systems.* Warsaw, Poland.

Kim T.H., Cho T.H., Moon Y.S. and Park S.H. (1999) Visual Inspection System for the Classification of Solder Joints. *Pattern Recognition.* Vol. 32, pp. 565 - 575.

Korel F., Luzuriaga D.A., Balaban M.O. (2001) Quality Evaluation of Raw and Cooked Catfish (Ictalurus Punctatus) Using Electronic Nose and Machine Vision. *Journal of Aquatic Food Product Technology.* Vol. 10, No. 1, pp. 3 - 13.

Krone A. and Teuber P. (1996) Applying WINROSA for Automatic Generation of Fuzzy Rule Bases. *EUFIT '96.* Aachen, pp. 929 - 932.

Kwak C., Ventura J.A. and Tofang-Sazi K. (2000) A Neural Network Approach for Defect Identification and Classification on Leather Fabric. *Journal of Intelligent Manufacturing.* Vol. 11, No. 5, pp. 485 - 499.

Lahajnar F., Bernard R., Pernus F. and Kovacic S. (2002) Machine Vision System for Inspecting Electric Plates. *Computers in Industry.* Vol. 47, No. 1, pp. 113 - 122.

Lavangnananda, K. (2001) Synergistic Genetic Algorithm for Inductive Learning (SynGAIL). *Proc. of the Int. ICSC Congress on Computational Intelligence Methods and Applications (CIMA 2001).* University of Wales Bangor, U.K. pp. 443 - 449.

Majumdar S. and Jayas D.S. (1999) Single-Kernel Mass Determination for Grain Inspection Using Machine Vision. *Applied Engineering in Agriculture.* Vol. 15, No. 4, pp. 357 - 362.

Melvyn L. and Richard J. (2000) Automated Inspection of Textured Ceramic Tiles. *Computers in Industry.* Vol. 43, No. 1, pp. 73 - 82.

Michalski R.S., Mozetic I., Hong J. and Lavrac N. (1986) The Multi-Purpose Incremental Learning System AQ15 and its Testing Application to Three Medical Domains. *Proc. 5th Int. Conf. on Artificial Intelligence.* Philadelphia, Pennsylvania. Morgan Kaufman, San Mateo, CA. pp. 1041-1045.

Moganti M., Ercal F., Dagli C.H. and Tsunekawa S. (1996) Automatic PCB Inspection Algorithms: A Survey. *Computer Vision and Image Understanding.* Vol. 63, No. 2, pp. 287 - 313.

Newman T.S. and Jain A.K. (1995) A Survey of Automated Visual Inspection. *Computer Vision, Graphics and Image Processing.* Vol. 61, No. 2, pp 231 - 262.

Nguyen H.T. and Walker E.A. (1999) *A First Course in Fuzzy Logic.* CRC Press. Boca Raton, FL.

Ni B., Paulsen M.R., Liao K. and Reid J.F. (1997) Design of an Automated Corn Kernel Inspection System for Machine Vision. *Trans. of the ASAE*. Vol. 40, No. 2, pp. 491 - 497.

Pearson T. and Toyofuku N. (2000) Automated Sorting of Pistachio Nuts with Closed Shells. *Applied Engineering in Agriculture*. Vol. 16, No. 1, pp. 91 -94.

Pham D.T. and Aksoy M.S. (1993) An Algorithm for Automatic Rule Induction. *Artificial Intelligence in Engineering*. Vol. 8, pp. 277 - 282.

Pham D.T. and Aksoy M.S. (1995a) RULES: A Simple Rule Extraction System. *Expert Systems Applications*. Vol. 8, pp. 59 - 65.

Pham D.T. and Aksoy M.S. (1995b) A New Algorithm for Inductive Learning. *Journal of Systems Engineering*. Vol. 5, pp. 115 - 122.

Pham D.T. and Alcock R.J. (1998) Automated Grading and Defect Detection: A Review. *Forest Products Journal*. Vol. 48, No. 4, pp. 34 - 42.

Pham D.T. and Chan A.B. (1998) Unsupervised Neural Networks for Control Chart Pattern Recognition. *CIRP Int. Seminar on Intelligent Computation in Manufacturing Engineering*. Capri (Naples), Italy, pp 343 - 350.

Pham D.T. and Dimov S.S. (1997a) An Efficient Algorithm for Automatic Knowledge Acquisition. *Pattern Recognition*. Vol. 30, No. 7, pp. 1137 - 1143.

Pham D.T. and Dimov S.S. (1997b) An Algorithm for Incremental Inductive Learning. *Proc. IMechE. Part B - Journal of Engineering Manufacture*. Vol. 211, No. 3, pp. 239 - 249.

Pham D.T. and Karaboga D. (2000) *Intelligent Optimisation Techniques: Genetic Algorithms, Tabu Search, Simulated Annealing and Neural Networks*. (2nd printing). Springer Verlag, Berlin and London.

Pham D.T. and Liu X. (1999) *Neural Networks for Identification, Prediction and Control*. (4th printing) Springer Verlag, Berlin and London.

Pham D.T. and Oztemel E. (1996) *Intelligent Quality Systems*. Springer Verlag, Berlin and London.

Pham D.T. and Pham P.T.N. (1988) Expert Systems in Mechanical and Manufacturing Engineering. *Int. Journal of Advanced Manufacturing Technology*. Special Issue on Knowledge Based Systems. Vol. 3, No. 3, pp. 3 - 21.

Pham D.T., Pham P.T.N. and Alcock R.J. (1998) Intelligent Manufacturing. In *Novel Intelligent Automation and Control Systems*. Vol. I. (ed. Pfeiffer J.) Papierflieger, Clausthal-Zellerfeld, Germany. pp. 3 - 18.

Quinlan J.R. (1983) Learning Efficient Classification Procedures and their Applications to Chess End Games. In *Machine Learning, an Artificial Intelligence Approach*. (eds. Michalski R.S., Carbonell J.G. and Mitchell T.M.) Morgan Kaufmann, San Mateo, CA. pp. 463 - 482.

Quinlan J.R. (1993) *C4.5: Programs for Machine Learning*. Morgan Kaufmann, San Mateo, CA.

Romanchik D. (2001) Vision System Checks Auto Axle Assemblies. *Image Processing Europe*. January/February. PennWell, Hounslow, UK. pp. 30 - 32.

Schalkoff R.J. (1997) *Artificial Neural Networks*. McGraw-Hill, New York.

Soini A. (2000) New Sensors for New Machine Vision Applications. *Sensor Review*. Vol. 20, No. 4, pp. 287 - 293.

Suga Y. and Ishii A. (1998) Application of Image Processing to Automatic Weld Inspection and Welding Process Control. *Int. Journal of the Japan Society for Precision Engineering*. Vol. 32, No. 2, pp. 81 - 84.

Sun D.W. (2000) Inspecting Pizza Topping Percentage and Distribution by a Computer Vision Method. *Journal of Food Engineering*. Vol. 44, No. 4, pp. 245 - 249.

Sylla C. (1993) Pairing Human and Machine Vision in Industrial Inspection Tasks. *Control Engineering Practice*. Vol. 1. pp. 171 - 182.

Tantaswadi P., Vilainatre J., Tamaree N. and Viraivan P. (1999) Machine Vision for Automated Visual Inspection of Cotton Quality in Textile Industries using Color Isodiscrimination Contour. *Computers and Industrial Engineering*. Vol. 37, No. 1-2, pp. 347 - 350.

Thomas A.D.H. and Rodd M.G. (1994) Knowledge-Based Inspection of Electric Lamp Caps. *Engineering Applications of Artificial Intelligence*. Vol. 7, No. 1, pp. 31 - 37.

Tolba A.S. and Abu-Rezeq A.N. (1997) A Self-Organizing Feature Map for Automated Visual Inspection of Textile Products. *Computers in Industry*. Vol. 32, No. 3, pp. 319 - 333.

Truchetet F., Jender H., Gorria P., Paindavoine M. and Ngo P.A. (1997) Automatic Surface Inspection of Metal Tubes by Artificial Vision. *Materials Evaluation.* Vol. 55, No. 4, pp. 497 - 503.

Tsai D.M. and Tseng C.F. (1999) Surface Roughness Classification for Castings. *Pattern Recognition.* Vol. 32, pp. 389 - 405.

Urena, R., Rodriguez, F., Berenguel M.A. (2001) Machine Vision System for Seeds Quality Evaluation using Fuzzy Logic. *Computers and Electronics in Agriculture.* Vol. 32, No. ER1, pp. 1 - 20.

UKIVA (2002) *Twenty-one Financial Justifications for using Machine Vision.* UK Industrial Vision Association. P.O. Box 25, Royston, Herts, UK.

Vivas C., Gomez Ortega J. and Vargas M. (1999) Automated Visual Quality Inspection of Printed Ceramic Dishes. In *Advances in Manufacturing* (ed. Tzafestas S.G.), Springer-Verlag, London. pp. 89 - 100.

Wang J., Harwood R.J. and Norton-Wayne L. (1997) The Use of Random-field Theory for Fault Detection in Automated Visual Inspection of Carpets. *Journal of the Textile Institute.* Vol. 88, No. 4, Part 1, pp. 476 - 487.

Wen Z.Q. and Tao Y. (1999) Building a Rule-based Machine-Vision System for Defect Inspection on Apple Sorting and Packing Lines. *Expert Systems with Applications.* Vol. 16, No. 3, pp. 307 - 313.

Wiltschi K., Pinz A. and Lindeberg T. (2000) An Automatic Assessment Scheme for Steel Quality Inspection. *Machine Vision and Applications.* Vol. 12, No. 3, pp. 113 - 128.

Yazdi H.R. and King T.G. (1998) Application of 'Vision in the Loop' for Inspection of Lace Fabric. *Real-time Imaging.* Vol. 4, No. 5, pp. 317 - 332.

Yu S.S., Cheng W.C., Chiang S.C. (1988) Printed Circuit Board Inspection System PI/1. *SPIE Automated Inspection High Speed Vision Architectures II.* Vol. 1004, pp. 126 - 134.

Zhou L.Y., Chalana V. and Kim Y. (1998) PC-based Machine Vision System for Real-time Computer-aided Potato Inspection. *Int. Journal of Imaging Systems and Technology.* Vol. 9, No. 6, pp. 423 - 433.

Problems

1. For the application of detecting brown marks on green apples, describe what might be carried out in the following processing stages: image acquisition, image enhancement, image segmentation, feature extraction and classification.

2. Discuss the following statement "vision inherently involves intelligence".

3. Write a simple expert system, containing if-then rules, to classify the following data.

	Height	Width	Type
Object 1	1	3	1
Object 2	2	4	1
Object 3	4	1	2
Object 4	3	2	2
Object 5	2	5	3
Object 6	1	8	3
Object 7	3	7	4
Object 8	4	6	4

4. Given two original chromosomes 110010 and 001101, show the resulting chromosome after a crossover of the middle two bits and a mutation of the least-significant bit.

5. Discuss the appropriateness of each of the following:
   - Expert systems for image acquisition
   - Neural networks for classification
   - Inductive learning for image enhancement

6. Consider a neuron with one input i, one weight w, one output o and no bias input. Additionally, the neuron uses the following activation function:

$$f(x) = 1 \text{ if } x \geq 0.5$$
$$f(x) = 0 \text{ if } x < 0.5$$

Thus, the output of the neuron is:

$$o = f(w.i)$$

Determine the range of values for w that would give the following input-output mapping:

x	o
1	0
2	0
3	1
4	1

7. Install the ProVision software from the attached CD onto your computer. Open the LABEL project. Read the description of the problem from the online notes. Select the following two options from the Test menu: *Breakpoint at end of cycle* and *Display test results*. Run the program and find the value of *Result* in the *I Digital Output* step.

8. In the *I Image Capture* step of the LABEL project, change the image to *label2*. Run the program again. What is the value of *Result* in this case?

Chapter 2

Image Acquisition and Enhancement

This chapter discusses the first two stages of the AVI process: image acquisition and image enhancement.

2.1 Image Acquisition

To inspect an object automatically, it is necessary to acquire data about the object. There are several sensors that can be used for this purpose. These include x-ray sensors, ultrasonic sensors and cameras. These are non-destructive test (NDT) or non-destructive evaluation (NDE) devices, as they do not damage the product during inspection.

X-rays are absorbed when they pass through a material. The degree of absorption depends upon the wavelength of the x-ray, the material of the object through which the rays are passing and the object's density. Thus, x-rays can be employed to inspect solid objects for internal defects. One example of the successful application of x-rays in inspection is their use to detect internally damaged pistachio nuts [Casasent et al., 1998]. With optical inspection, worm damaged nuts could not be seen and so a large number of good nuts were rejected to compensate for this. However, x-ray inspection has two main problems. First, it produces large amounts of data, the processing of which can create difficulties for real-time application. Second, the equipment is expensive to purchase.

Ultrasonic waves are high frequency sound waves inaudible to the human ear. One method of using ultrasonic waves for inspection is to transmit them towards the object and then analyse the reflected signal. The transmitter can also be deployed as a receiver. Ultrasonic testing methods have been adopted for many tasks, such as pipeline [Kopp, 2001] and turbine inspection [Quirk, 2001].

This book concentrates upon automated inspection using cameras. One disadvantage of visual inspection when compared with other methods is that it cannot detect internal defects. Another problem is that it can become confused by factors such as dirt and shadows. However, visual inspection is gaining increasing interest in the inspection of industrial products and there are several reasons for this. First, a large amount of research has already been carried out into image processing. Second, humans normally perform inspection with their eyes, motivating the use of cameras. Third, the price of video equipment and image processing software is falling rapidly due to their widespread application.

Digital images consist of a number of pixels (picture elements). For example, an image of size 512x512 pixels has a width of 512 pixels and a height of 512 pixels, making a total of 262,144 pixels in the image. Each pixel has a value to indicate the intensity of light at that point. In a binary image, pixels have a value of zero (black) or one (white). In a grey-scale image, each pixel has a grey level. If there are Z grey-levels in the image then each pixel has a grey level from zero (black) to $Z-1$ (white). Pixel values in a grey-scale image represent the greyness of the pixel. Pixels with low grey levels are dark grey whilst those with high grey levels are light grey. A colour image can be thought of as three grey-scale images. The intensity of the three primary colours (red, green and blue) is stored in three images. However, sometimes compression techniques are used for colour images to reduce their storage requirements.

The two main types of camera are called line-scan and area-scan cameras. An area-scan camera is designed for acquiring images of objects that are not in motion. Area-scan cameras are used for objects on a conveyor belt in two ways. First, the object can be momentarily stopped so that a blur-free image can be taken. Second, strobe lighting can be used, where the light flashes for a fraction of a second. This requires synchronisation of the camera and lighting. A line-scan camera acquires an image consisting of just one line. If the object is moving with respect to the camera, sequential lines are put together to make a complete image.

Figure 2.1 illustrates the process of image acquisition and image enhancement. First, the object is positioned in front of the camera and an image is obtained. Second, the image can be enhanced to create an improved image.

When capturing an image, it is very important to ensure the highest possible image quality. Without a good quality image, it is difficult to inspect the product accurately. Time spent improving image acquisition saves efforts expended on developing complicated image processing algorithms. To acquire high-quality images, it is necessary to pay careful consideration to the lighting and object placement. The main reason for this is that ambient lighting can seriously reduce image quality.   Different lighting configurations are appropriate for detecting different objects.

When using cameras to capture images, there are also other factors which need to be taken into account, for example, whether colour vision should be employed and the resolution of the images that is required. The image quality can also be reduced by conditions in industrial environments such as dust and dirt.

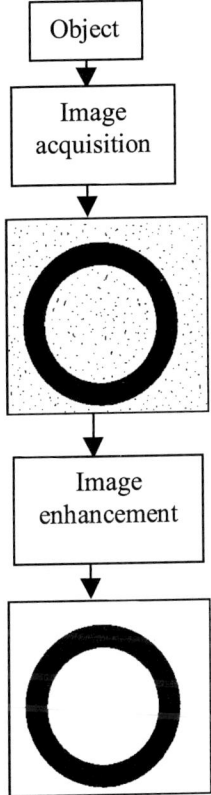

Figure 2.1 Image acquisition and enhancement

2.1.1 Lighting System Design

Lighting is one of the most important factors in obtaining high quality images. Ambient lighting is subject to large variations and so it is not a recommended lighting method. Typically, the inspection area is surrounded by a black box and controlled light sources are used. The current importance of lighting in inspection system design is highlighted by Braggins [2000], who states that machine vision illumination is itself becoming an industry. In the beginning of AVI, the inspection system designer would purchase lights from an electrical shop but now there are companies available to offer specialised advice on machine vision lighting systems.

Figure 2.2 shows a categorisation of lighting techniques. The three main approaches are called frontlighting, structured lighting and backlighting.

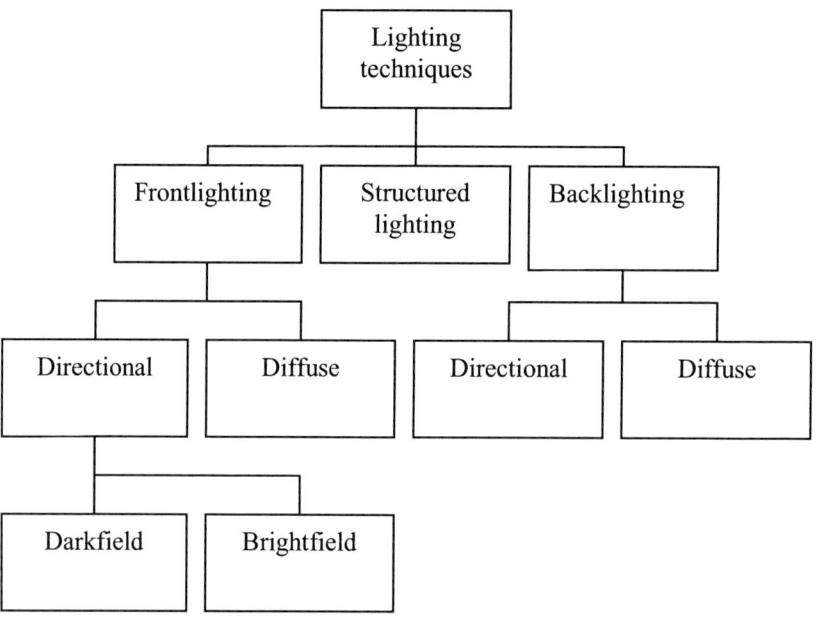

Figure 2.2 Lighting techniques

Frontlighting is lighting from above the object, which is utilised when the surface finish of an object is important. It can be further divided into directional and diffuse frontlighting.

An example of a directional frontlighting device is a spotlight.   Directional frontlighting may be either brightfield or darkfield. Both techniques use an inclined light source. In brightfield illumination, the camera is placed at the angle of the reflected light. This is illustrated in Figure 2.3. Brightfield illumination is used to detect 3D surface features. The angle of incidence of the light source depends upon the height of the surface features. Darkfield illumination is shown in Figure 2.4. In this type of illumination, a directional light source is shone at a low angle of incidence onto the surface of the object. This type of illumination is used to detect surface scratches.   If the angle of incidence is reduced to 0° then side lighting is obtained. This is useful, for example, for inspecting labels on transparent objects.

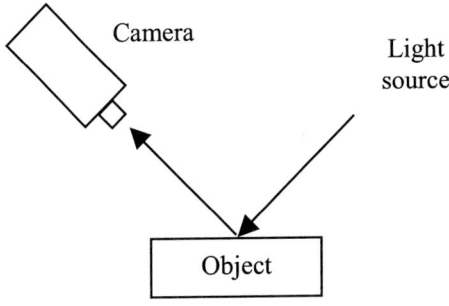

Figure 2.3 Brightfield illumination [Burke, 1996]

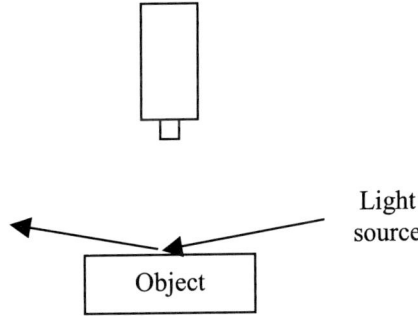

Figure 2.4 Darkfield illumination [Burke, 1996]

A problem with directional frontlighting is glinting, which is caused when the light is reflected off a shiny surface, such as metal or glass. One method to avoid glinting is to utilise polarised light, produced by passing the light through a polarising filter.

Another useful technique is given by Batchelor et al. [1985]. The method is called omni-directional illumination and is illustrated in Figure 2.5. The idea is to avoid the use of direct illumination. A light is shone upwards onto a white surface that reflects the light back down onto the object being inspected. Direct illumination from the light to the object is shielded using a small wall. A problem that has been identified with this set up is that a reflection of the camera lens may be seen on the object. To overcome this, frontlighting passed through a beamsplitter can be applied from a position next to the camera to give diffuse illumination [Braggins, 2000].

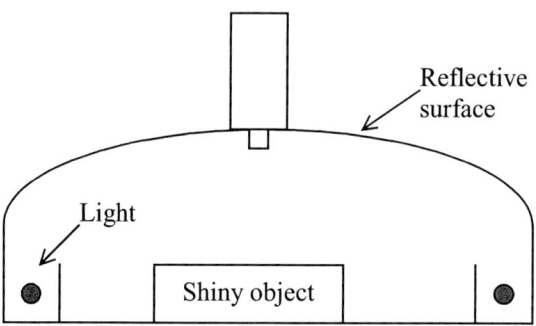

Figure 2.5 Omni-directional illumination [Batchelor et al., 1985]

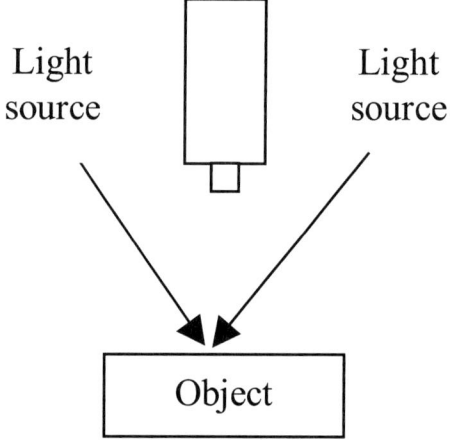

Figure 2.6 Diffuse frontlighting

Diffuse frontlighting is simple and commonly adopted. It is illustrated in Figure 2.6. One method of providing even diffuse frontlighting is to use a fluorescent ring light, where the lens of the camera is placed through the centre of the ring.

Lighting from behind the object is called backlighting (Figure 2.7). It produces an image of high contrast, where the background and holes appear very bright and the object is very dark. The object appears as a silhouette on a bright background. Backlighting is employed to inspect object shape but cannot be used to inspect surface features. Like frontlighting, backlighting may also be directional or diffuse. To produce diffuse light, a diffuser is placed in front of the light source. Directional backlighting is good for producing sharp edges, whereas diffuse backlighting is better for increasing image contrast.

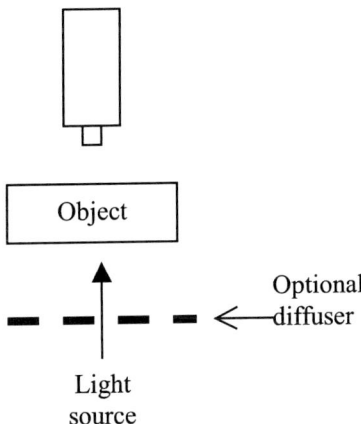

Figure 2.7 Backlighting

Structured lighting yields 3D information about the object. The idea is to project a structured light pattern, such as stripes, onto the object. The deflection of the stripes at any point indicates the height of the object at that point. When no object is present, the stripes are completely straight. As the height of the object onto which the stripes are projected increases, so too does the deflection of the stripes.

The position of the lighting used to acquire the image has a large effect on the quality of the image. Figures 2.8 to 2.10 give examples of the effect of different lighting positions when viewing the same object. Figure 2.8 shows the image when only the front light is used. Figure 2.9 displays the effect of employing only a light from behind the object. This image would be useful for measuring the circularity of the object or checking the hole in the centre. Figure 2.10 illustrates the effect of combining both lighting methods. The result is similar to using frontlighting only but the contrast between the seal and the hole is better.

Figure 2.8 Front light only

Figure 2.9 Back light only

Figure 2.10 Combined lighting

2.1.2 Light Sources

There are many types of light source available for AVI applications. These include [Burke, 1996]:

- Ambient light;
- Incandescent tungsten lamps;
- Quartz-halogen lamps;
- Arc and gas discharge lamps;
- Lasers;
- LEDs;
- Spectral light sources.

Ambient light includes natural sunlight and light from the surroundings. In a manufacturing environment, surrounding light consists mainly of fluorescent ceiling lights. Advantages of ambient light are that it is readily available and cheap. However, the major disadvantage is that it is uncontrollable by the vision system engineer. The intensity of daylight varies considerably over the course of a day and from day to day. This has a very significant effect on the image acquisition process. Therefore, it is recommended that the inspection area be shielded from ambient light using a black box. A more expensive alternative in the long run is to include light sources that are so strong that the effect of ambient light is negligible.

Incandescent lamps are very popular for AVI applications. These bulbs are widely available and low cost. They operate by passing an electric current through a tungsten filament surrounded by inert gases, such as argon and nitrogen. Disadvantages of these bulbs include low efficiency and large source area. Also, due to filament thinning and bulb blackening, the intensity of the light deteriorates significantly over the lifetime of the lamp. The amount of intensity that a bulb can lose before it needs to be changed is an important factor for AVI systems.

To improve the efficiency of tungsten bulbs, the temperature of the filament can be increased. However, this leads to reduced bulb lifetime. To overcome this problem, quartz-halogen, or tungsten-halogen, bulbs were developed. In these bulbs, a small amount of halogen material is added into the bulb to enable higher filament temperatures to be reached. Halogen materials include iodine and bromine. Quartz glass is used to surround the gas to cope with the higher temperatures. Compared to tungsten bulbs, quartz-halogen bulbs have much longer lifetimes and higher operating efficiency.

Bulbs based on filaments are not efficient because a large amount of energy is wasted as heat. Arc-discharge lamps are much more efficient than incandescent lamps. Fluorescent lights are a common type of arc-discharge lamp, which consist of a tube filled with a small amount of mercury and a mixture of gases. To initiate the arc discharge in the light, a large starting voltage is required. Fluorescent lights are suitable for illuminating large areas. Several fluorescent lights can be used to provide a relatively even illumination area. For AVI applications, it is important to note that the intensity of illumination of a fluorescent light is not as strong at the ends of the tube. Compared with incandescent lamps, fluorescent lamps give more light in the blue part of the colour spectrum. It is stated in Braggins [2000] that the illumination of domestic fluorescent lights varies with the frequency of the power source. To avoid this problem for inspection, lights with long-persistence phosphor are required. Someji et al. [1998] inspected castings using fluorescent light. Uniform illumination was obtained by using several fluorescent ring lights.

Lasers provide lighting that is intense and highly directional. Basically, a laser is amplified coherent light, generated by an intense beam of photons, which can be focussed onto a small area. A problem with laser is that it is not an efficient form of lighting. However, laser light is ideal for producing structured lighting as it can be focussed extremely precisely. Two main types of lasers are helium-neon and carbon-dioxide lasers. Helium-neon lasers are very common. Helium and neon gases are mixed in a glass tube and when the helium atoms are excited, photons are allowed to escape at one end of the tube. Carbon-dioxide lasers are very powerful and therefore more suitable for industrial cutting tasks than for AVI applications. Oyeleye and Lehtihet [1999] inspected solder joints on PCBs by illuminating the joint with a laser line. The presence and amount of solder was detected by measuring the deflection of this line. In the inspection of wood planks, Stojanovic et al. [2001] employed a laser-line generator to provide structured light. Using triangulation theory, range information was extracted from the acquired images to determine shape deformations in the plank.

LEDs (light emitting diodes) are very small sources of light that produce nearly monochromatic intense light. Several LEDs mounted in an array can provide sufficient light for inspection. Titus [2001] highlighted the increased use of LEDs for industrial inspection for three reasons. First, they are low cost compared to other lighting methods. Second, they have a relatively long lifetime. Third, LEDs can be turned on and off very quickly. This means that they can be employed in applications where strobe lighting is needed or where different patterns of lighting are required at different times during the inspection cycle. LED lights can be arranged to create spotlights, ring lights, diffuse lighting and back lighting [Wilson, 2001]. They have now become the preferred source of illumination in the machine vision industry.

Depending on the object being inspected, different parts of the light spectrum may be suitable. Thus, infrared or ultraviolet illumination may be beneficial. Most objects emit infrared radiation naturally, so infrared illumination is seldom required. Infrared techniques are suitable to inspect items that have internal defects. It has been utilised to inspect objects such as pipes [Maldague, 1999], soya beans [Nakamura, 1999], almonds [Pearson, 1999] and apples [Wen and Tao, 1999]. Outside the area of AVI, infrared sensing is used widely by the police and military.

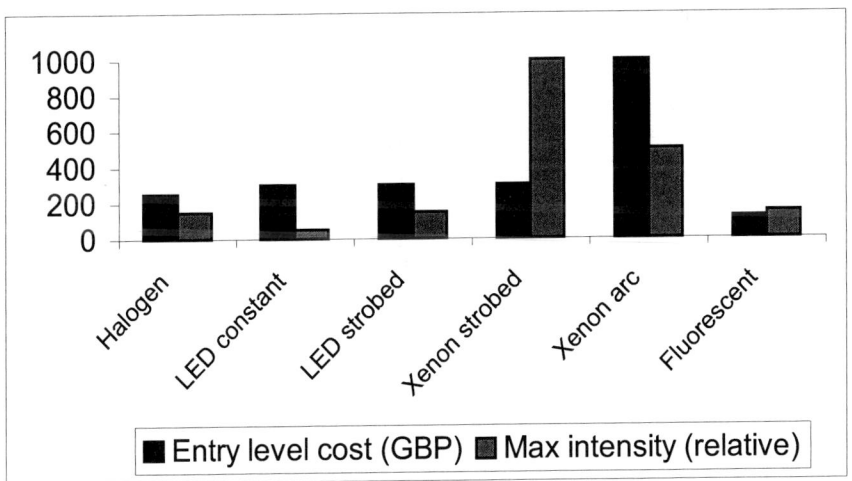

Figure 2.11 Cost and intensity of lighting sources

Ultraviolet light is useful in cases when the object being inspected fluoresces. Chen and Tretiak [1992] utilised fluorescence imaging to inspect integrated circuit lead bonds. It was not expected that fluorescent light would be useful in this application but it was tried when other types of illumination did not give images of sufficient quality.

A comparison of the various lighting methods, both in terms of intensity levels and costs, is found in Braggins [2000] (Figure 2.11). The information displayed for halogen lamps is for direct illumination using fibre-optic guides. Xenon arc lamps are rarely adopted in AVI tasks, mainly due to their high cost.

2.1.3 Choosing Optimum Lighting

A significant problem in AVI is how to find the optimum lighting for a given inspection task. Harding [2001] describes lighting as an art rather than a science. However, to help new developers of inspection systems, lighting system design must be formalised.

Van Dommelen [1996] states that it is necessary to consider the characteristics of the part to be inspected before thinking of the lighting technique to employ. The three most important optical characteristics of an object are its surface reflectance, geometry and colour:

- **Reflectance** can be *specular, diffuse* or *directional*. Specular surfaces are smooth and reflect light in a specific direction. Diffuse surfaces are rough and return light in many directions. Surfaces that are directionally reflective return light at different angles depending upon the angle of incidence of the light source.

- The **geometry** of an object may be *flat, curved* or *prismatic*. It is much easier to obtain uniform illumination for a flat part than for a curved part. Prismatic parts contain sharp edges, which can create shadows.

- **Colour** may be useful in segmenting parts of an object if different areas of the part have different colours. For example, red lighting can be employed to emphasise red areas of the part. This can be achieved using a red filter on the camera or by adopting illumination that transmits light with the red wavelength.

A comprehensive range of AVI lighting techniques was described by Batchelor et al. [1985]. Many techniques are given, such as how to view holes and pipes, looking around corners and detecting cracks. To guide users in the choice of appropriate lighting methods, Batchelor [2002] has developed a computer program called the

Lighting Advisor, which can be accessed on the Internet. The Lighting Advisor contains an index, describing some 180 lighting techniques. The techniques covered include combining several views of an object; measuring wall thickness and observing bubbles in clear fluid. Information about each technique is divided into a number of sections, including objective, equipment needed, typical applications, references, camera and lens type, object details (such as motion, geometry and surface finish) and additional remarks. Other researchers have produced lighting expert systems. Novini [1994] developed a lighting and optics expert system that advises the user on appropriate configuration of the image acquisition system. A similar system was designed by West [1993] to aid camera lens selection. The system queries the user about their requirements and then recommends a suitable lens from its knowledge base.

Table 2.1 describes general lighting methods and the types of object for which they are most appropriate.

Lighting technique	Application areas
Frontlighting - directional - darkfield	3D surface features
Frontlighting - directional - brightfield	Surface scratches
Structured lighting	3D objects
Backlighting	2D object shape, dimensions and holes
Omni-directional lighting	Shiny objects

Table 2.1 Lighting methods and their most suitable applications

A lighting testbed system called ALIS (All-Purpose Lighting Investigation System) is available from the lighting suppliers Dolan-Jenner [2002]. ALIS contains several light sources, which can be moved around to generate various lighting configurations.

To illustrate the process of lighting selection, Van Dommelen [1996] gave examples of lighting for three applications:

- **White lettering on black cases.** The lettering has a high contrast from the background and the casing has a dull surface. Two lighting techniques are suggested. First, a fluorescent ring light could be utilised to give uniform front illumination. Second, polarised light could be employed to reduce any problems with glinting.

- **Characters etched into a silicon wafer.** The surface of the wafer is metallic and reflective, so diffuse frontlighting would not produce a good image. The suggested lighting methods are darkfield or brightfield illumination, which would give a good contrast between the letters and the background. This can be achieved using two banks of LED lights directed at an angle onto the wafer.

• **Checking solder on PCBs**. The solder joint may be unfilled or "solder bridges" may be formed. For this application, two types of lighting are required. The solder joints are grey, reflective and have various shapes. Diffuse front illumination was found to be appropriate for them. For the detection of solder bridges, an image was acquired using directional polarised light. This image was then subtracted from the image acquired with diffuse illumination. The resulting image then has a high contrast between the bridges and the background.

For additional help in the selection of appropriate lighting, there are now many AVI lighting experts who work in companies that supply light sources for machine vision. A list of suppliers is available online at [TMWorld, 2002; MVOnline, 2002; VSDesign, 2002].

2.1.4 Other Image Acquisition Issues

A common problem in AVI is that the light sources give uneven lighting over the inspection area. One technique to cope with this was given by Sawchuk [1977]. First, an image is acquired of a piece of card with an homogeneous shade. Second, the image is smoothed and, finally, it is compared with the acquired image using image subtraction.

Karjalainen and Lappalainen [2000] stated that it is better to optimise the lighting rather than to try to compensate for uneven lighting at a later stage. Their work involved determining the optimum position of lamps used for acquiring images with a line-scan camera. The lights were placed above the conveyor belt at a fixed angle. Thus, the problem was how to find the optimum positions and intensities for each of the lamps to produce even illumination along a line. The technique developed was based on regularisation theory combined with least-squares estimation. Promising results were found in three case studies.

In a manufacturing environment, there is the problem of dust and dirt reducing the image quality. Dust may accumulate on the object and also on the lens of the camera. Kim and Koivo [1994] found that the presence of dust in the inspection area did degrade the quality of the acquired images in their inspection system. They suggested the inclusion of a fan to remove the dust.

Choosing the resolution of the images correctly is also important. If too low a resolution is selected then small defects cannot be detected. If too high a resolution is chosen then the amount of information to be processed and the execution time increase. Funck et al. [1990] found that adopting the correct resolution improved the detection accuracy for certain defect types by more than 20%.

On conveyor belts, the objects are moving and therefore, images of the objects may become blurred. As already mentioned, a method to overcome this problem is to use a strobe light and synchronised camera to reduce the blurring effect. An example of an inspection system that included strobe lighting was developed by Ni et al. [1997] to inspect corn kernels. The system was designed to eliminate blur caused by the movement of the kernels. The strobe was synchronised with the blanking period of the camera by using control circuitry.

An example of an image of an object to be inspected is shown in Figure 2.12. The image is of a car engine seal, consisting of 256 grey levels. It will be used throughout the book to illustrate various image-processing operations and will be referred to as the example image.

Figure 2.12 Example image

2.1.5 3D Object Inspection

As the popularity of AVI increases, applications are moving from 2D inspection to 3D inspection [Hardin, 2000]. Lee et al. [2000] describe the three most popular methods for obtaining 3D data as stereo vision, "time of flight" techniques and structured lighting. In a stereo-based system, computational matching algorithms are employed to determine corresponding points in two images. Problems with this are that there could be too many or too few points to correlate. The matching problem is also not simple to solve. Time of flight methods, such as lasers, give a large amount of accurate data, however, they are very costly. Structured lighting is gaining popularity for 3D data gathering, due to its relative simplicity and low cost. As previously mentioned, a light pattern is shone onto a 3D object and the deflections in the pattern are measured. The patterns include a point, stripe, several stripes or a grid. Commonly, a multiple-stripe pattern is adopted. Figure 2.13 shows how multiple stripes can be used to obtain height information from a rectangular object.

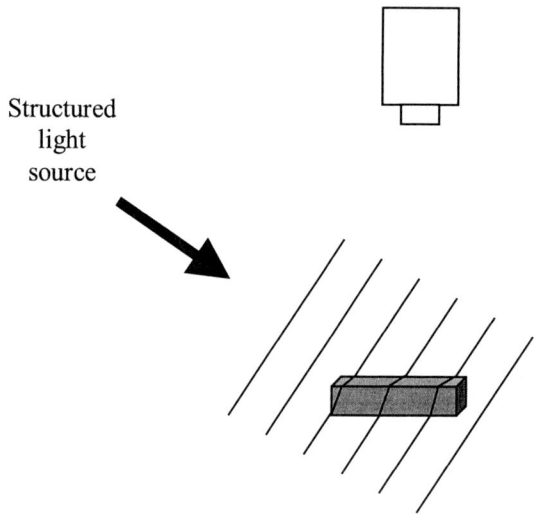

Structured
light
source

Figure 2.13 Structured lighting for 3D data gathering

Lee et al. [2000] applied structured lighting for grading herring roe fish. Initially, a 2D-inspection system was developed but it was found that certain deformities on the herring, such as dents, could not be detected. Height information derived from the 3D images was able to improve weight estimation of the fish. The system consists of five stages. First, an image is acquired under structured lighting, consisting of multiple stripes. Second, the peaks in the stripes are found. Third, each individual stripe is located using neighbourhood tracking. Fourth, the correlation is found between the deflections in the stripes and the real-world co-ordinates. Finally, the information found is employed either for shape interpretation or for volume estimation.

Another AVI application in which 3D information is particularly beneficial is the inspection of surface-mount-technology (SMT) assemblies in the electronics industry [Guerra and Villalobos, 2001]. Aspects of SMT assemblies that make them difficult to identify with 2D images are the small component size, reflections and colour.

To improve upon the performance achievable using visual methods for 3D data capture, there are many alternative techniques. Peng and Loftus [2001] highlight co-ordinate-measuring machines (CMMs) and laser methods as common alternatives to vision-based methods. CMMs operate using a touch probe, which contacts the surface of the object at specified points. Laser methods gather data at a distance from the object. These methods can give very accurate results but their limiting factors are their cost and slow execution times.

2.2 Image Enhancement

Image enhancement techniques may be applied to the acquired image to make it more suitable for the subsequent processing stages. The type of image enhancement chosen depends upon the segmentation to follow. Two of the main reasons for enhancing an image are to remove noise and to improve certain features in the image. Enhancement of specific image features is designed with the overall application in mind.

The main objectives of image enhancement are conflicting. First, it tries to remove noise from the image, which has the effect of making edges less sharp. Second, it attempts to sharpen the edges of objects, which introduces noise pixels into the image. Thus, a balance must be found where the edges are sufficiently sharp and there is an acceptably low amount of noise in the image.

A technique for enhancing the contrast of an image is histogram equalisation. An image histogram shows the number of pixels in the image having each grey level. The idea of histogram equalisation is to ensure that a roughly equal number of pixels in the image have each individual grey level. After equalisation, the histogram is flat and thus does not contain peaks or troughs. Therefore, histogram equalisation is not useful if histogram-based segmentation techniques are to be used. Histogram equalisation can improve the contrast of an image so that it is better for a human viewer but it is not normally beneficial as a pre-processing stage for segmentation.

Chen et al. [1995] proposed a method based on fuzzy set operations for specifying the shape of the image histogram. The technique can be employed to generate a brighter image when the acquired image is dark. This makes the image clearer to a human viewer but, again, would not enhance the image for histogram-based segmentation techniques.

Techniques for image enhancement that could be used in AVI include smoothing and morphology. To explain these methods, convolution filters will be described first, as they are an integral part of many image enhancement operations. Figure 2.14 shows a representation of an original image of size 4x4 pixels, a convolution filter of size 3x3 and an output image of size 4x4. The original image X is combined with the convolution filter C and a new image Y is obtained. Then, image Y is the enhanced version of the input image. Pixel y_{22} in the output image is the corresponding enhanced pixel value for pixel x_{22} in the original image.

x_{11}	x_{12}	x_{13}	x_{14}
x_{21}	x_{22}	x_{23}	x_{24}
x_{31}	x_{32}	x_{33}	x_{34}
x_{41}	x_{42}	x_{43}	x_{44}

Original image

c_{11}	c_{12}	c_{13}
c_{21}	c_{22}	c_{23}
c_{31}	c_{32}	c_{33}

Convolution filter

y_{11}	y_{12}	y_{13}	y_{14}
y_{21}	y_{22}	y_{23}	y_{24}
y_{31}	y_{32}	y_{33}	y_{34}
y_{41}	y_{42}	y_{43}	y_{44}

Output image

Figure 2.14 Application of convolution filters

For example, to calculate the value of pixel y_{22} in the output image, the following equation is employed:

$$y_{22} = \frac{x_{11}c_{11} + x_{12}c_{12} + x_{13}c_{13} + x_{21}c_{21} + x_{22}c_{22} + x_{23}c_{23} + x_{31}c_{31} + x_{32}c_{32} + x_{33}c_{33}}{c_{11} + c_{12} + c_{13} + c_{21} + c_{22} + c_{23} + c_{31} + c_{32} + c_{33}}$$

(2.1)

Figure 2.15 illustrates three simple examples of an original image, a convolution filter and the resulting output image. Values cannot be calculated for the pixels at the border of the image as the convolution filter will not fit into the image at this point. Thus, a 3x3 filter reduces the size of the image by one pixel on all sides. For large images, this effect is not serious. If the result of applying the convolution filter gives a negative value then these may be stored as zeros if the frame buffers cannot store negative pixel values.

2.2.1 Low-pass Filtering

One of the most common methods of enhancement is image smoothing, known as low-pass filtering. For this, convolution filters are used. As mentioned previously, a problem with smoothing is that it makes edges less sharp in addition to removing noise pixels. Three common methods of smoothing are averaging, Gaussian smoothing and median filtering.

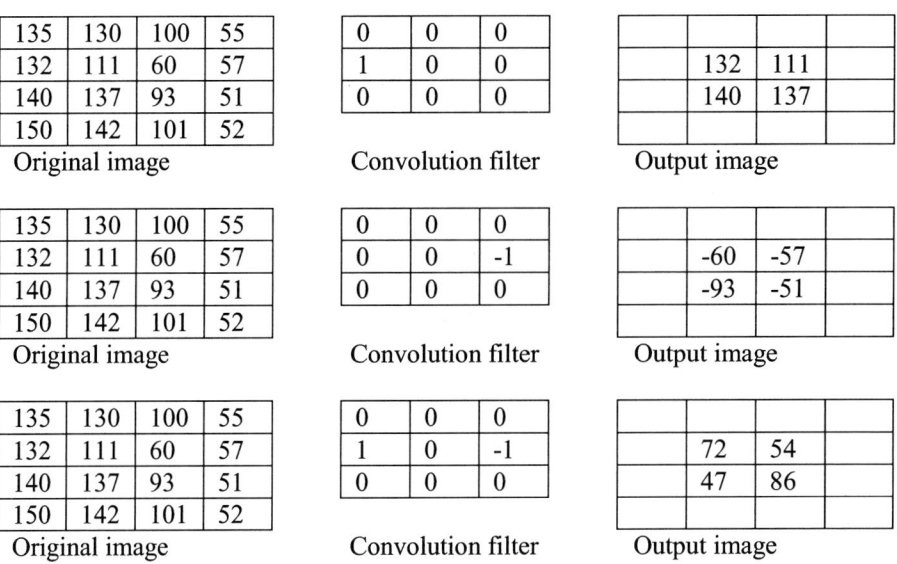

Figure 2.15 Examples of using convolution filters

The simplest type of smoothing is mathematical averaging. Figure 2.16 shows how this operation is represented in convolution filter format. The filter contains nine ones because each grey level is multiplied by one before being summed. Next, the resulting sum is divided by nine. Larger filters, such as 5x5 and 7x7, are also popular for smoothing.

1	1	1
1	1	1
1	1	1

Figure 2.16 3x3 averaging convolution filter

Figure 2.17 shows the result of performing 7x7 averaging on the example image. It can be seen that the metal section surrounding the central hole is blurred.

Figure 2.17 Result of averaging the example image

Gaussian smoothing operates in a similar manner to averaging smoothing but uses different convolution filters. The convolution masks for 3x3 and 5x5 smoothing are shown in Figure 2.18. In each case, the resulting pixel values are divided by 256. Gaussian smoothing has a smaller smoothing effect than averaging as it gives a higher influence to the centre pixel. However, for a human observer, Gaussian smoothing has a very similar effect to averaging smoothing.

The median filter operates by replacing the grey level of a pixel with the median value of pixels in a neighbourhood surrounding the pixel. Compared with simple averaging, the median operation is better able to remove noisy pixels whilst maintaining edge information. However, the sorting operation required means that it is more time consuming.

25	30	25
30	36	30
25	30	25

5	8	9	8	5
8	13	16	13	8
9	16	19	16	9
8	13	16	13	8
5	8	9	8	5

Figure 2.18 3x3 and 5x5 Gaussian convolution filters

2.2.2 Morphology

Morphology is a standard image-processing tool. It can be performed on grey-scale or binary images. Structuring elements are required, which can be of arbitrary shape but the most often employed are square elements of size 3x3, 5x5 or 7x7. To explain morphology, it is first necessary to describe the image processing operations of *erosion* and *dilation*. Erosion is a reduction operation that is implemented with the 3x3 minimum convolution filter. This filter replaces the grey level of a pixel with the lowest grey level in the 3x3 neighbourhood. A pixel in a binary image will become black if any of its neighbouring pixels are black. Dilation is an operation that grows the object and can be implemented using the 3x3 maximum convolution filter. This filter replaces the grey level of a pixel with the highest grey level in a 3x3 neighbourhood centred on the pixel itself. In a binary image, this operation turns a pixel white if any adjacent pixel is white.

The effects of 3x3 dilation and erosion on a single white pixel in a binary image are shown in Figure 2.19. Erosion causes the pixel to disappear whereas dilation enlarges the pixel to a 3x3 square. The process of applying one or more erosion operations followed by an equal number of dilation operations, with the same structuring element, is called *opening*.

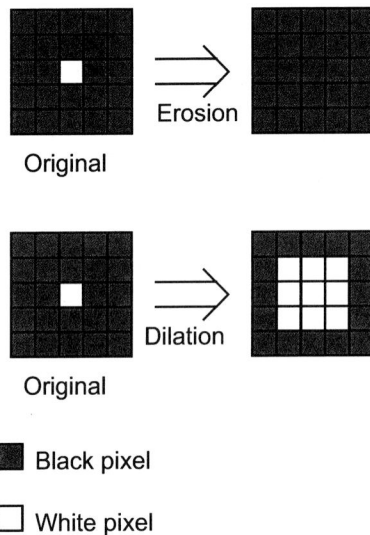

Figure 2.19   Erosion and dilation operations

Figures 2.20 and 2.21 show the effect of performing erosion and dilation, respectively, on the example image. These operations were implemented using 7x7 minimum and maximum convolution filters. In Figure 2.20, the metal section surrounding the central hole cannot be seen whereas in Figure 2.21, this area is emphasized.

Figure 2.22 shows how the application of one opening operation will remove a solitary pixel but would not remove a 3x3 square. Opening results in a smoothing of the object boundary but the object preserves its basic size (unless it is removed). By using different structuring elements, for example a horizontal line instead of a square, only objects of a certain shape will be removed. Figure 2.23 shows the effect of applying one 7x7 opening operation to the example image.

Figure 2.20 Erosion performed on the example image

Figure 2.21 Dilation performed on the example image

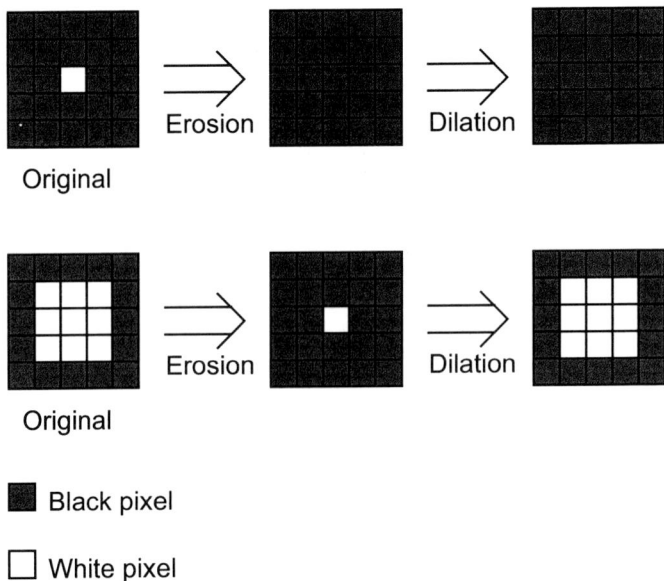

Original

Erosion

Dilation

Original

Erosion

Dilation

■ Black pixel

□ White pixel

Figure 2.22 Morphological opening

Morphological closing is also implemented using the operations of erosion and dilation but in the reverse order to opening. *Closing* consists of one or more dilation operations followed by an equal number of erosion operations, with the same structuring element. The name is derived from the fact that small holes in light objects are closed. Figure 2.24 shows the effect of applying one 7x7 closing operation to the example image.

Figure 2.25 shows how one closing operation, with a 3x3 square structuring element, can be used on a binary image to join together two white pixels separated by a space one pixel wide. When two nearby objects are dilated, they will touch each other. If erosion is subsequently applied, the link between the objects often remains.

Figure 2.23 Morphological opening on example image

Figure 2.24 Morphological closing on example image

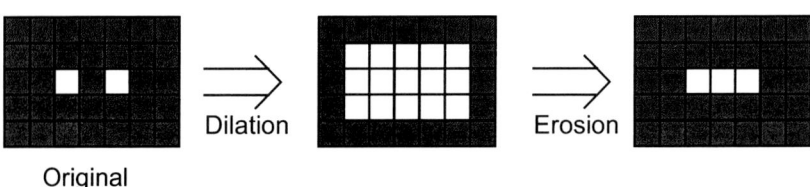

Dilation   Erosion

Original

Figure 2.25 Morphological closing

2.2.3 Artificial Intelligence for Image Enhancement

Much research has been performed in image enhancement and several researchers have employed artificial intelligence techniques in the process.

Research has been carried out in the authors' laboratory into image enhancement using AI techniques [Pham and Bayro-Corrochano, 1992; Bayro-Corrochano, 1993]. A technique was developed for edge and line enhancement, based on iterative relaxation methods [Rosenfeld and Kak, 1982]. It involved several processing stages:

1. An edge detection method is employed to find edges in the image.

2. A pixel is chosen at random.

3. Four convolution masks are applied to the image to obtain the degree of evidence of an edge at the chosen pixel.

4. The generated evidence value is compared with a threshold. If it is larger than the threshold then the evidence and intensity of the pixel are increased. Otherwise, they are decreased.

5. Return to stage (2) until the whole image is covered.

6. Return to stage (2) unless the stopping criterion has been satisfied. The energy of the image was employed for this instead of the usual entropy or mean squared error.

The reinforcement of image areas with a strong evidence and influence of pixels on their neighbours is considered similar to the operation of the Kohonen SOFM.

Shih et al. [1992] utilised an ART-based neural network for image enhancement. In the proposed network, a two-layer ART model was embedded into a four-layer neural network. Image enhancement filters were applied that operated as region and contour detectors. The technique was able to fill gaps in the image and reduce noise.

Russo [1998] gave an overview of where fuzzy logic has been employed in image enhancement. Fuzzy enhancement techniques are divided into two categories, called direct and indirect approaches. Direct approaches are termed FIRE (Fuzzy Inference Ruled by Else-action) systems. A FIRE system changes the grey level of pixels with the aid of fuzzy rules. The conditions of the fuzzy rules contain statements about the grey level values of the neighbours of the pixel being analysed. If no rule is satisfied then the grey level of the pixel is not changed. This is called the Else-action.

To explain the FIRE technique in more detail, an example will be given. Figure 2.26 shows a 3x3 area of an image. Pixel x_0 is the pixel for which noise removal is to take place. The enhanced value of x_0 will be $x_0 + \Delta y$.

	x_1	
x_2	x_0	x_3
	x_4	

Figure 2.26 Pixel values used for FIRE system input

The inputs to the fuzzy rules are Δx_1, Δx_2, Δx_3 and Δx_4, where:

$$\Delta x_1 = x_1 - x_0 \tag{2.2}$$

$$\Delta x_2 = x_2 - x_0 \tag{2.3}$$

$$\Delta x_3 = x_3 - x_0 \tag{2.4}$$

$$\Delta x_4 = x_4 - x_0 \tag{2.5}$$

An example of a rule in a FIRE system is:

IF Δx_1 is POSITIVE BIG and Δx_2 is POSITIVE BIG and Δx_3 is POSITIVE BIG
    THEN Δy is POSITIVE BIG

In the indirect approach to fuzzy image enhancement, Δy is normally obtained using a network of weighted connections similar to a neural network. An example of such a system is given in Figure 2.27.

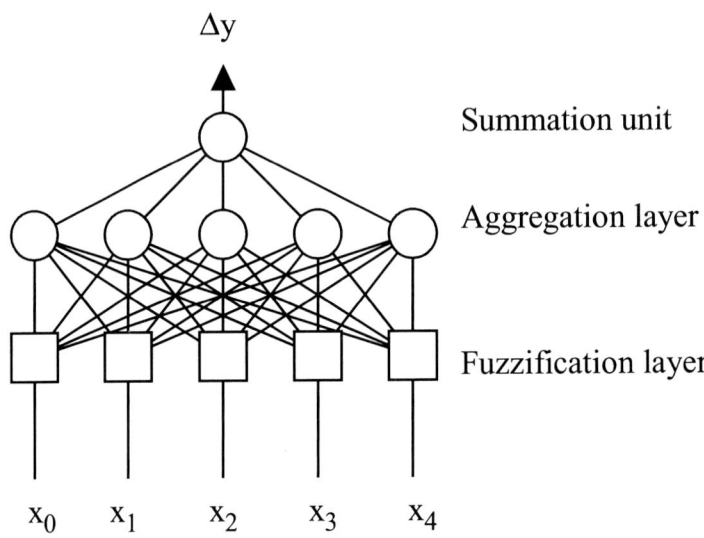

Figure 2.27 Example of a fuzzy image enhancement network

Choi and Krishnapuram [1997] used fuzzy logic for selecting which image enhancement function to employ. It was stated that one problem with the FIRE method is that a large number of fuzzy rules are required to deal with all possible pixel grey-level values. To overcome this, Choi and Krishnapuram proposed that the output of the fuzzy system should not be Δy but the desired type of filtering. Three filters were proposed, one for images with low noise levels, one for images with high noise levels and a third where the noise level could not be predetermined. The execution time of the method was found to be similar to that of the median filter.

Work has also been carried out into employing genetic algorithms in image enhancement. Pal et al. [1994] adopted GAs to search for the correct image enhancement function to employ in a given situation. Poli and Cagnoni [1997] stated that image enhancement is normally performed with some kind of convolution filter and proposed that these convolution filters could be evolved using GAs. One problem of this approach is how to find a suitable fitness function. The solution proposed was to use a symbolic-regression-like function. The error was calculated by comparing the output of the image after filtering with the desired output image. As each image contains a large number of pixels, the total processing time required by the GA could be very large. Therefore, the image was sampled and the GA executed with the reduced data set. A second possible problem with the approach is that the desired output image is not normally available.

2.3 Discussion

Two factors are important for image enhancement in AVI systems. The first is performance. Image enhancement techniques should improve noticeably the quality of the image ready for segmentation. The approach of most AVI system engineers is to avoid using image enhancement techniques altogether. It is preferred to spend more time and effort on improving lighting techniques than on devising complex image enhancement processes.

The second important factor in using image enhancement in AVI is speed. In manufacturing environments, items are produced rapidly. Very little time is available to inspect each item and so image enhancement methods should be designed to be as simple as possible.

2.4 Summary

Selection of the optimum lighting method plays a key role in successful inspection. However, lighting system design has often been described as an art form and much work still needs to be done to formalise it.

It is possible to employ image-enhancement techniques, such as smoothing or morphology, to improve the quality of acquired images. However, in practical applications, image enhancement is not normally used as it is considered better to spend the effort on improving the image acquisition stage.

References

Batchelor B.G. (2002) *Lighting Advisor*. Dept. of Computer Science, Cardiff University, UK. bruce.cs.cf.ac.uk/bruce/index.html

Batchelor B.G., Hill D.A. and Hodgson D.C. (1985) *Automated Visual Inspection*. IFS, Bedford, UK.

Bayro-Corrochano E.J. (1993) *Artificial Intelligence Techniques for Machine Vision*. PhD thesis, School of Engineering, Cardiff University, UK.

Braggins D. (2000) Illumination for Machine Vision. *Sensor Review*. Vol. 20, No. 1, pp. 20 - 23.

Burke M.W. (1996) *Image Acquisition: Handbook of Machine Vision Engineering, Volume 1*. Chapman and Hall, London.

Casasent D.A., Sipe M.A., Schatzki T.F., Keagy P.M. and Lee L.C. (1998) Neural Net Classification of X-ray Pistachio Nut Data. *Food Science and Technology*. Vol. 31, No. 2, pp. 122 - 128.

Chen B.T., Chen Y.S. and Hsu W.H. (1995) Automatic Histogram Specification Based on Fuzzy Set Operations for Image Enhancement. *IEEE Signal Processing Letters*. Vol. 2, No. 2, pp. 37 - 40.

Chen J. and Tretiak O.J. (1992) Fluorescence Imaging for Machine Vision. *Applied Optics*. Vol. 31, No. 11, pp. 1871 - 1877.

Choi Y.S. and Krishnapuram R. (1997) A Robust Approach to Image Enhancement Based on Fuzzy Logic. *IEEE Trans. on Image Processing*. Vol. 6, No. 6, pp. 808 - 825.

Dolan-Jenner (2002) *ALIS - All Purpose Lighting Investigation System*. Lawrence, MA. www.dolan-jenner.com

Funck J.W., Brunner C.C. and Butler D.A. (1990)   Softwood Veneer Defect Detection Using Machine Vision. *Proc. Symp. on Process Control / Production Management of Wood Products*. Forest Products Research Society, Madison, WI, pp. 113 - 120.

Guerra E. and Villalobos J.R. (2001)   A Three-Dimensional Automated Visual Inspection System for SMT Assembly. *Computers and Industrial Engineering*. Vol. 40, No. 1, pp. 175 - 190.

Hardin W. (2000) New Approaches in 3D Applications. *Machine Vision Online.* www.machinevisiononline.org.

Harding K. (2001) The Art of Lighting Science. *Machine Vision Online.* www.machinevisiononline.org.

Karjalainen P.A. and Lappalainen T. (2000) Optimization of Illumination Profiles in Line-Scan Camera Systems. *Measurement Science Technology.* Vol. 11, No. 9, pp. 1301 - 1306.

Kim C.W. and Koivo A.J. (1994) Hierarchical Classification of Surface Defects on Dusty Wood Boards. *Pattern Recognition Letters.* Vol. 15, No. 7, pp. 713 - 721.

Kopp F. (2001) High-resolution Sizing Capabilities Required for Automated Ultrasonic Inspection of Offshore Pipeline and Catenary Riser Welds. *Pipes and Pipelines International.* Vol. 46, No. 3, pp. 25 - 41.

Lee M.F., De Silva C.W., Croft E.A. and Wu Q.M. (2000) Machine Vision System for Curved Surface Inspection. *Machine Vision and Applications.* Vol. 12, No. 4, pp. 177 - 188.

Maldague X. (1999) Pipe Inspection by Infrared Thermography. *Materials Evaluation.* Vol. 57, No. 9, pp. 899 - 902.

MVOnline (2002) Buyer's Guide: Lighting Equipment. *Machine Vision Online.* www.machinevisiononline.org.

Nakamura O., Kobayashi M. and Kawata S. (1999) Nondestructive Inspection of Phaseolus Coccineus L Soya Beans by Use of Near-infrared Lasers. *Applied Optics.* Vol. 38, No. 12, pp. 2724 - 2727.

Ni B., Paulsen M.R., Liao K. and Reid J.F. (1997) Design of an Automated Corn Kernel Inspection System for Machine Vision. *Trans. of the ASAE.* Vol. 40, No. 2, pp. 491 - 497.

Novini A. (1994) The Lighting and Optics Expert System for Machine Vision. In *Selected Papers on Machine Vision Systems.* (eds. Batchelor B.G. and Whelan P.F.) SPIE Optical Engineering. Bellingham WA. pp. 167 - 171.

Oyeleye O. and Lehtihet E.A. (1999) Automatic Visual Inspection of Surface Mount Solder Joint Defects. *Int. Journal of Production Research.* Vol. 37, No. 6, pp. 1217 - 1242.

Pal S.K., Bhandari D. and Kundu M.K. (1994) Genetic Algorithms for Optimal Image Enhancement. *Pattern Recognition Letters*. Vol. 15, No. 3, pp. 261 - 271.

Pearson T.C. (1999) Use of Near Infrared Transmittance to Automatically Detect Almonds with Concealed Damage. *Food Science and Technology*. Vol. 32, No. 2, pp. 73 - 78.

Peng Q. and Loftus M. (2001) Using Image Processing Based on Neural Networks in Reverse Engineering. *Machine Tools and Manufacture*. Vol. 41, pp. 625 - 640.

Pham D.T. and Bayro-Corrochano E.J. (1992) Neural Computing for Noise Filtering, Edge Detection and Signature Extraction. *Journal of Systems Engineering*. No. 2, pp. 111 - 122.

Poli R. and Cagnoni S. (1997) Evolution of Psuedo-colouring Algorithms for Image Enhancement with Interactive Genetic Programming. *Proc. 2nd Int. Conf. on Genetic Programming*. Stanford. Morgan Kaufmann. pp. 269 - 277.

Quirk K. (2001) The Application of Automated Ultrasonic Equipment to the Inspection of Turbine Generators. *Insight*. Vol. 43, No. 9, pp. 590 - 592.

Rosenfeld A. and Kak A.C. (1982) *Digital Picture Processing*. (2nd ed.). Academic Press, New York.

Russo F. (1998) Recent Advances in Fuzzy Techniques for Image Enhancement. *IEEE Trans. on Instrumentation and Measurement*. Vol. 47, No. 6, pp. 1428 - 1434.

Sawchuk A.A. (1977) Real-Time Correction of Intensity Non-Linearities in Imaging Systems. *IEEE Trans. on Computers*. Vol. 26, No. 1, pp. 34 - 39.

Shih F.Y., Moh J.L. and Chang F.C. (1992) A New ART-Based Neural Architecture for Pattern Classification and Image Enhancement with Prior Knowledge. *Pattern Recognition*. Vol. 25, No. 5, pp. 533 - 542.

Someji T., Yoshimura T. and Akiyama N. (1998) Development of Automatic Surface Inspection System of Castings. *Int. Journal of the Japan Society for Precision Engineering*. Vol. 32, No. 4, pp. 278 - 283.

Stojanovic R., Papadopoulos G., Mitropoulos P., Alcock R.J. and Djurovic I. (2001) An Approach for Automated Inspection of Wood Boards. *Int. Conf. on Image Processing*. IEEE Signal Processing Society. Thessaloniki, Greece. pp. 798 - 801.

Titus J. (2001) Lights Make Machine Vision Shine. *Test and Measurement World.* Vol. 21, No. 7, pp. 33 - 36. www.tmworld.com

TMWorld (2002) Online Buyer's Guide: Lighting, Machine Vision. *Test and Measurement World.* www.tmworld.com

Van Dommelen C.H. (1996) Choose the Right Lighting for Inspection. *Test and Measurement World.* October, pp. 53 - 58. www.tmworld.com

VSDesign (2002) Product Guide: Image Sources, Optics and Illumination. *Vision System Design.* vsd.pennnet.com

Wen Z.Q. and Tao Y. (1999) Building a Rule-based Machine-vision System for Defect Inspection on Apple Sorting and Packing Lines. *Expert Systems with Applications.* Vol. 16, No. 3, pp. 307 - 313.

West P. (1993) Development of an Expert System to Aid Machine Vision Application. *Proc. Int. Conf. on Robotics and Vision Automation.* pp. 21/ 1 - 7.

Wilson A. (2001) LED Lighting Devices Add Smart Controls. *Vision Systems Design.* August 2001. vsd.pennnet.com

Problems

1. What lighting arrangement would be suitable for the following applications:
- Detecting scratches on a casting;
- Measuring features on a coin;
- Checking the dimensions of a machined part.

2. Calculate the resulting output image when the following convolution filter is applied to the given image grey levels:

97	103	110	115
105	107	121	123
107	114	134	140
110	117	156	160

0	2	0
0	0	0
0	-2	0

Original image   Convolution filter

3. What would happen to the following image, containing just black and white pixels, after dilation using a 3x3 structuring element?

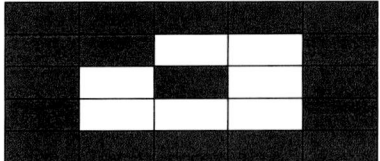

4. Apply erosion to the image produced after dilation has been performed on the image of Problem 3. Show the resulting image. What is this operation called?

5. How would the following image appear after application of the averaging filter?

97	103	110	115
105	107	121	123
107	114	134	140
110	117	156	160

6. Consider a transparent shampoo bottle. Describe its reflectance, geometry and colour characteristics. Determine suitable lighting for reading the text on the bottle front.

7. In ProVision, the function to implement convolution filters is called *W Filter*. The PACK project utilises this function three times. Run the program for all five test images and find the result of the digital output for all five cases.

8. Choose one of the three *W Filter* steps in the PACK project and open the associated dialog box by double clicking on it. Change the filter type from Sobel to low pass. Press the button *Test* and then the button *Display* to see the effect of the operation on the selected image window. Change the mask size to 7x7 and describe what effect this has on the output window.

Chapter 3

Segmentation

Integral to the process of AVI is segmentation, which decomposes an image into constituent objects or regions. For some AVI applications, segmentation is simply the specification of a fixed region of interest in the image from which features can be derived. For complex inspection tasks, such as surface inspection, segmentation is the most complicated part of AVI [Pham and Alcock, 1996]. Humans are able to segment images easily using information such as edges, colour and background knowledge. However, computers have problems in segmenting all but simple scenes. One possible reason why humans are so much better than computers at segmentation is that a large amount of background knowledge is used. Programming such background information into computers is a time-consuming and difficult task.

Figure 3.1 illustrates the segmentation process. The acquired image, with or without enhancement, is presented to the segmentation module. Employing some image processing functions, any objects of interest in the image are separated from the background and each other. Then, the background and objects are made into contrasting shades, usually black on white or vice versa.

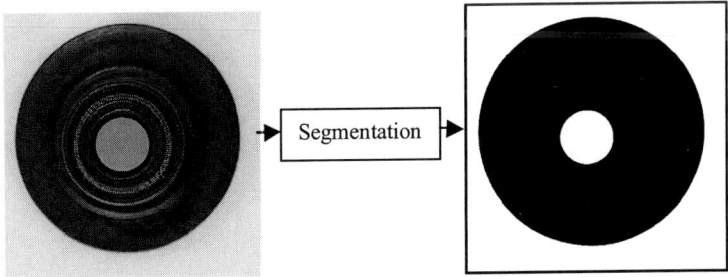

Figure 3.1 Segmentation process

Many different techniques are available for segmentation. The most common are edge detection, thresholding and region growing. Other techniques include split-and-merge, windows-based subdivision, template matching, profiling and AI-based segmentation.

3.1 Edge Detection

The basic assumption of edge detection is that regions can be separated because of the high differential of grey levels at the edge of each region. A well-known edge detection method is that by Sobel, which can be implemented by using eight 3x3 convolution filters [Gonzalez and Woods, 1992]. There are two horizontal, two vertical and four diagonal filters, as shown in Figure 3.2. The result of applying each filter is placed into a new image. For the overall result, these images need to be combined. The easiest method of doing this is to add all the windows together.

-1	0	1
-2	0	2
-1	0	1

1	0	-1
2	0	-2
1	0	-1

1	2	1
0	0	0
-1	-2	-1

-1	-2	-1
0	0	0
1	2	1

2	1	0
1	0	-1
0	-1	-2

0	1	2
-1	0	1
-2	-1	0

0	-1	-2
1	0	-1
2	1	0

-2	-1	0
-1	0	1
0	1	2

Figure 3.2 Sobel convolution filters

Figure 3.3 displays the result of applying the four horizontal and vertical Sobel edge detectors (shown in the top row of Figure 3.2) to the example image. Figure 3.4 displays the result of summing the contents of these four windows. The outside edge of the seal is detected strongly because of the relatively large difference in intensity between the seal and the background.

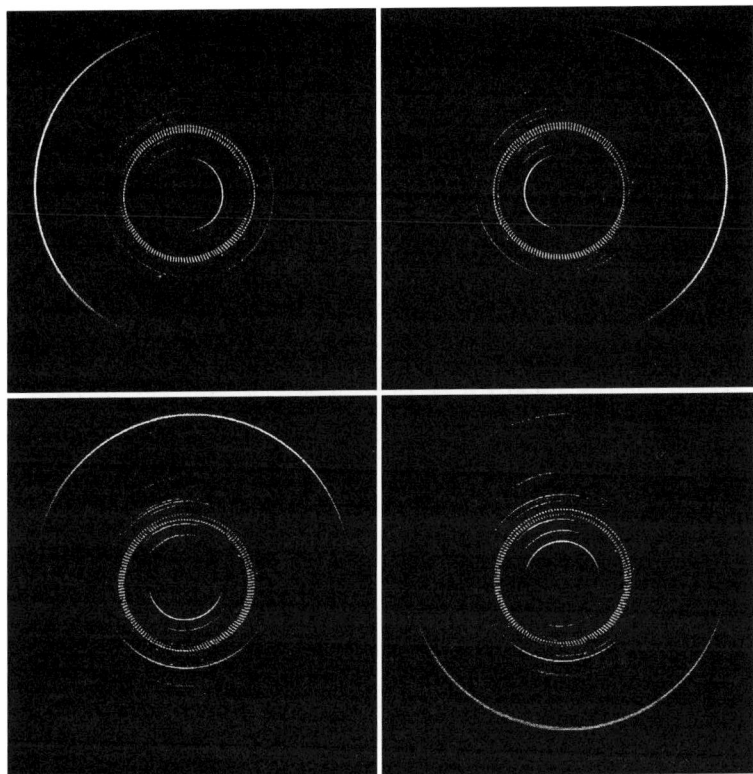

Figure 3.3 Result of applying Sobel filters to the example image

Several other edge detectors have been developed. The Prewitt edge detector is similar to the Sobel detector (Figure 3.5). Another common edge detector is the Roberts detector, which is an edge detector of size 2x2 instead of 3x3. Figure 3.6 shows the Roberts detector using 3x3 convolution filters, as convolution filters require a central pixel.

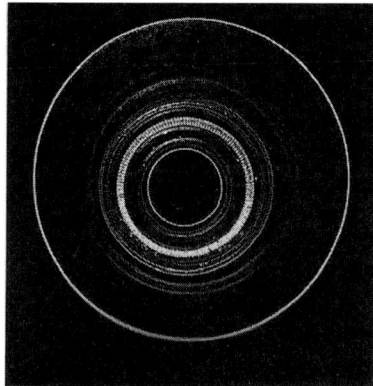

Figure 3.4 Result of adding the four Sobel edge-detected images

-1	0	1
-1	0	1
-1	0	1

1	0	-1
1	0	-1
1	0	-1

1	1	1
0	0	0
-1	-1	-1

-1	-1	-1
0	0	0
1	1	1

1	1	0
1	0	-1
0	-1	-1

0	1	1
-1	0	1
-1	-1	0

0	-1	-1
1	0	-1
1	1	0

-1	-1	0
-1	0	1
0	1	1

Figure 3.5 Prewitt edge detector

0	0	0
0	1	0
0	0	-1

0	0	0
0	-1	0
0	0	1

0	0	0
0	0	1
0	-1	0

0	0	0
0	0	-1
0	1	0

Figure 3.6 Roberts edge detector

Edge detection methods continue to be developed [Acton and Mukherjee, 2000; Sundaram, 1999; Zhu et al., 1999].   De Santis and Sinisgalli [1999] adopted a Bayesian approach to edge detection.   First, a Bayesian procedure is used to estimate the model parameters.   Second, likelihood ratio statistics are calculated as a hypothesis test to determine whether pixels belong to edges.

Hou and Kuo [1997] developed an edge detection method for image inspection that gave a better performance than the Sobel detector.   The algorithm consisted of three stages: image binarisation, image contraction and image subtraction. The generated edges were one pixel wide and continuous and therefore of good quality.

Artificial intelligence techniques have been applied to edge detection. These include neural networks, fuzzy logic and genetic algorithms. Pham and Bayro-Corrochano [1992] used neural networks for edge detection. The MLP neural network was chosen, as this is the most established neural network. Figure 3.7 shows the structure of the simple single layer MLP initially implemented for the problem. The inputs to the network were the grey levels of a pixel and its immediate neighbours (P1-P9), as would be used as inputs to a 3x3 convolution filter. The outputs of the network were the evidence of the centre pixel being an edge and the orientation of that edge. It was found that the task of learning for the network was very difficult, particularly as a large number of noisy images must be learnt. To evaluate the performance of the new edge detector, Abdou and Pratt's edge evaluation method was utilised [Abdou and Pratt, 1979]. The performance of the neural edge detector was not better than that of the Sobel edge detector.

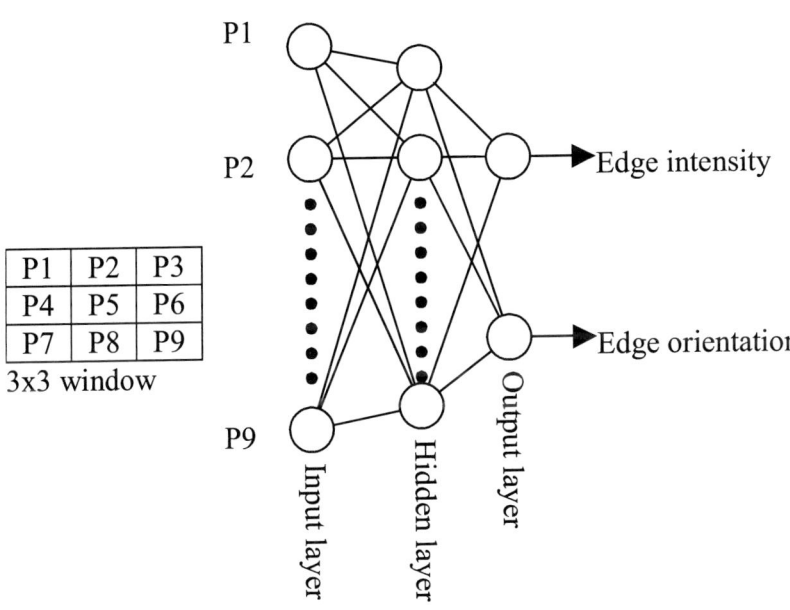

Figure 3.7 Neural edge detector

Pham and Bayro-Corrochano [1992] developed a new system where the pixel values were split into eight orientations (Figure 3.8). The system, which comprised eight MLP modules, had a synergistic structure; more details on synergy will be given in Chapter 5. The MAXNET took the largest value of all eight MLP modules and gave this as the final output. The new system gave a better performance than the Sobel detector, when the signal-to-noise (SNR) ratio of the image was between 2 and 20. Thus, the synergistic neural edge detector is particularly suited to detecting edges in noisy images.

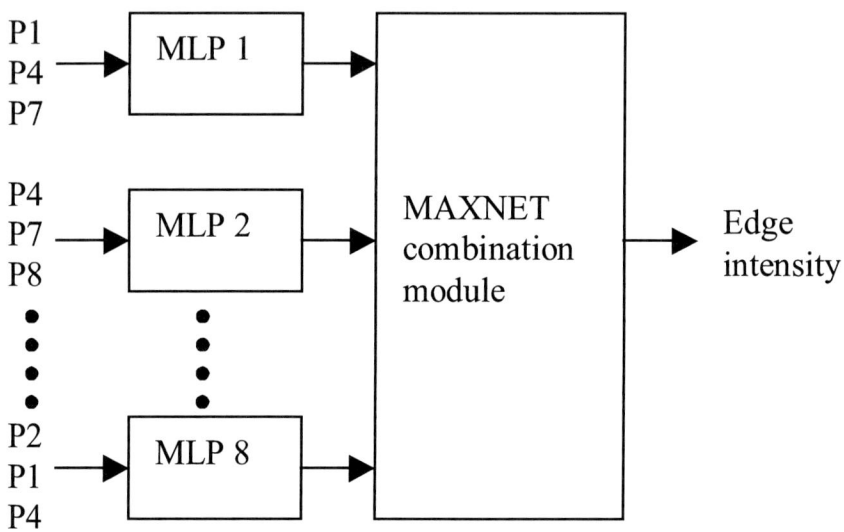

Figure 3.8 Synergistic neural edge detector

Murtovaara et al. [1996] searched for edge contours with a fuzzy logic method. The membership function for each contour type was generated and updated using historical contour data. According to their results, the technique filtered out fast contour changes. Russo [1998] also employed fuzzy logic in edge detection. Experimental results showed that the technique performed better than traditional edge detection methods in terms of suppression of noise and detection of image details.

Edge detection was performed using genetic algorithms by Gudmundsson et al. [1998]. The GA was compared with simulated annealing using the Pratt measure. The edges found by the GA were thin and continuous, which gave good values for the Pratt measure. Further experiments were proposed using different fitness functions and genetic operators.

There are several problems with edge detection. First, the edges are often found to be incomplete and then some method of edge linking is required to join the parts of the edges. This is not a simple task and also increases the time required for the method. Second, many erroneous edges are generated. It is then not trivial to determine that these do not represent actual object boundaries. Wen and Xia [1999] developed a technique to verify located edges for inspection purposes. Edge quality criteria were calculated, such as edge strength, edge length and neighbourhood grey-level distribution. Based on these criteria, the edge is either kept or deleted.

A method developed to overcome the problem of broken edges is called edge relaxation. If a pixel is between two parts of an edge and is also very similar to an edge pixel then it is likely that this pixel is also part of the edge. Such pixels can be added to the edge. Then, iteratively, the criterion of what is an edge is gradually relaxed and pixels near to edges can be tested to see if they match the new relaxed criterion.

Whilst one edge detector may be better than another for a given task, in real industrial applications, the choice of edge detector should not greatly affect the overall performance of the system. According to Batchelor and Whelan [1997], if the choice of edge detector makes a significant difference then it is likely that there is a problem with another part of the system, probably the lighting. Pavlidis [1992] found that the difference between the performance of two edge detectors on the same image was relatively small when compared with the performance of an edge detector on two different images. Thus, current edge detection methods cannot be applied to all types of images.

To facilitate edge detection, the lighting should be positioned so that edges are given a higher contrast. Das and Mukhopadhyay [1998] attempted to find edges of objects using optical methods with a lens-based optical set-up.

3.2 Thresholding

The simplest and fastest of segmentation methods is thresholding and this has been used widely. Thresholding is based on the idea that different objects or regions in the image have significantly different grey levels. Thresholds are usually determined from the grey-level histogram of an image.

Figure 3.9 shows the histogram for the example image. The image consists of grey levels from 0 to 255. It can be seen that the histogram contains two large peaks. The leftmost peak takes values of grey levels roughly between 10 and 80. This corresponds to most of the pixels that represent the seal. The rightmost peak takes values roughly between 180 and 220, representing most of the background pixels. Segmentation of this image can be performed by thresholding at grey level values somewhere between the two main peaks. Figure 3.10 shows the result of thresholding the example image at a value of 120.

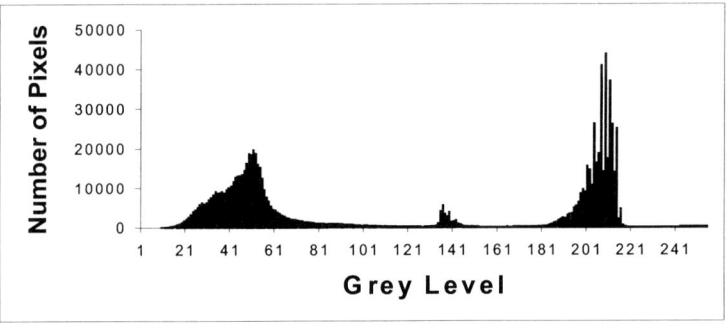

Figure 3.9 Histogram for the example image

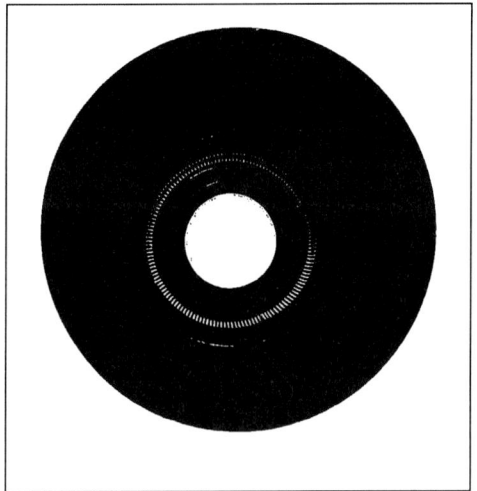

Figure 3.10 Result of thresholding the example image

Thresholding may be *global* (the same threshold is applied to the whole image) or *local* (different thresholds are applied to different parts of the image). It may also be *fixed* (the threshold value is the same for every image) or *adaptive* (the threshold is determined differently for each image or region). Global thresholding is very fast and simple but there are two major problems. First, it is often difficult to determine, automatically or even manually, the optimal threshold. Second, only in very simple images can a threshold segment all objects correctly. To perform local thresholding, an approach is to divide the image into square sub-regions and calculate the threshold independently for each one. However, there may be some discontinuities between edges of different regions.

There are two basic types of histogram: *unimodal* and *multimodal*. A unimodal histogram has one central peak whereas a multimodal histogram has more than one peak. A special case of the multimodal histogram is the *bimodal* histogram, which has two peaks. Frequently, the number of peaks in the histogram has some correlation with the actual objects in the image.

Determining the threshold from bimodal histograms has been widely studied and researchers have reported that the threshold should be placed at the valley in the histogram to segment the image [Gonzalez and Woods, 1992]. Figure 3.11 gives an example of an idealised bimodal histogram. The two peaks should correspond to two areas of the image that have different shades. The optimum threshold is marked at the place where the valley between the two peaks occurs. It should be noted that there is some overlap between the two distributions and so no threshold would be able to segment the two areas perfectly. Also, the histogram shown is idealised. In real situations, the histogram shape would be noisier and smoothing would be required to remove false minima.

Finding the best threshold from unimodal histograms is a more difficult task and researchers do not agree on any one technique. It is possible to determine the threshold on the basis of the average and standard deviation grey levels of the image histogram [Pham and Alcock, 1996]. Figure 3.12 shows a histogram obtained from the image of a wood board. It can be seen that the distribution formed is almost a normal distribution. From statistical theory, it is known that in such a distribution, roughly 95% of points lie within two standard deviations from the mean and approximately 99% of pixels are found within three standard deviations from the mean. Therefore, a suitable threshold was considered to be equal to the mean minus three standard deviations. Another possibility is to choose the threshold according to the valley points and inflexion points of the histogram [Otsu, 1979]. A survey of thresholding techniques can be found in Sahoo et al. [1988].

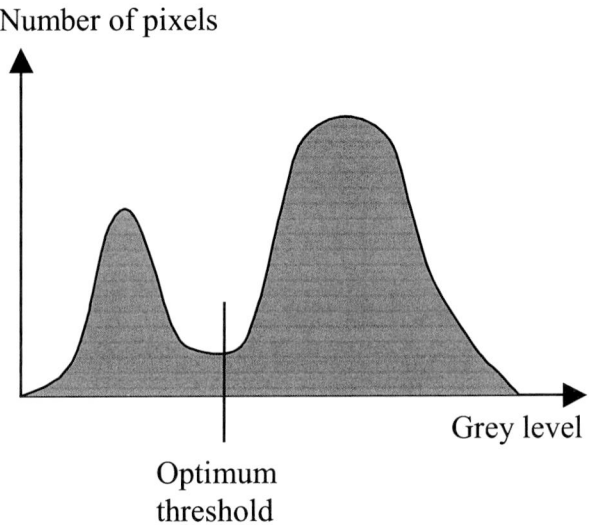

Figure 3.11 Bimodal histogram

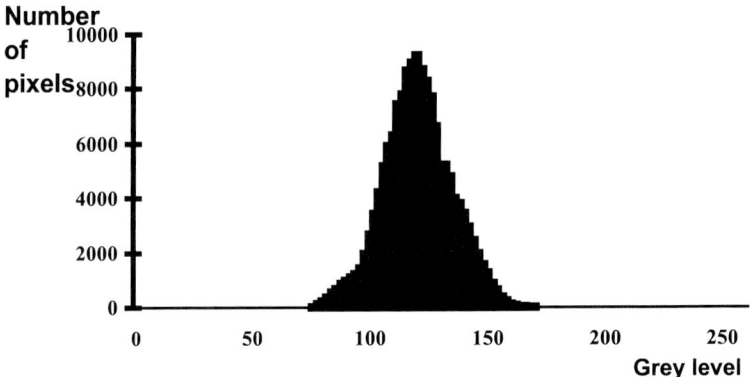

Figure 3.12 Unimodal histogram

It is possible to use a MLP neural network to determine the optimal threshold [Kothari et al., 1991]. One way to do this is to have a network with Z inputs, where Z is the number of grey levels, and one output. The inputs would be the values of the histogram and the output would be the best threshold for that histogram. As in all neural network applications, a large and representative training set would be required.

There have been numerous applications of thresholding in AVI. Heinemann et al. [1994] used thresholding for the inspection of mushrooms. It was found that the overall performance of the system depended upon the features that were extracted from the images. Other applications of thresholding include its use in the inspection of lace [Sanby et al., 1995], wood [Pham and Alcock, 1996] and print on plastic containers [Truchetet and Cholley, 1997].

For colour images, there are three histograms, one each for the red, green and blue components. The simplest method of thresholding is to determine the threshold of each histogram independently. Tuning of each threshold can then be performed based on the other two thresholds. In the inspection of potatoes, Zhou et al. [1998] found that the colour green was particularly useful, as bad potatoes change colour.

In many cases, global thresholding does not work well. Even manual selection of the threshold often cannot segment the image accurately. One possible cause for this is that the illumination may vary over the image and different thresholds may be needed in different areas. This is performed with local thresholding. As mentioned before, the image is divided into sub-windows and the threshold is calculated separately for each sub-window. Another way of solving the problem is to redesign the acquisition set-up or the lighting so that thresholding can be employed more effectively. The second approach is adopted in many AVI systems.

3.3 Region Growing

Region growing is a bottom-up method of segmentation. The idea of region growing is to take a seed pixel, grow it and thus create quasi-homogenous regions. The growing operation involves checking whether neighbouring pixels are similar. If they are, the pixels are placed into the same region. Pixels are considered to be part of the same region if they match given criteria. To implement region growing, it is necessary to decide upon appropriate seeds and growing criteria for the task. One of the simplest criteria is to join pixels according to the similarity of their grey levels.

There are two main problems with region growing. First, if the threshold is set too low then many regions are found. Second, if the threshold is set too high then the whole image is segmented into one large region. These problems with region growing can be illustrated with the simple example of Figure 3.13 [Gonzalez and

Woods, 1992]. Figure 3.13a displays the grey levels of pixels in a 4x4 region. Figure 3.13b shows that two regions are created when the threshold equals two. If a threshold of three is chosen, then the two regions are merged into one. This is because there are two pixels at the bottom of the image that are separated by a grey level of three (Figure 3.13c).

1	3	7	7
2	2	8	8
1	3	7	7
3	4	7	8

(a)

1	3	7	7
2	2	8	8
1	3	7	7
3	4	7	8

(b)

1	3	7	7
2	2	8	8
1	3	7	7
3	4	7	8

(c)

Figure 3.13 Region growing using different thresholds

A further problem with region growing is that some regions may not be labelled at all at the end of segmentation. To overcome this, new seeds can be placed in these areas and region growing can recommence.

If too many regions are created, extra processing is needed to merge them. An improvement to just utilising pixel grey level for similarity is to compare the shape of the evolving region with *a-priori* information.

Region growing is sensitive to which seeds are chosen initially. The final segmented image depends strongly on where the initial seeds are positioned. Ideally, the seeds should be placed at the centre of objects but this is only possible if *a-priori* information is known about the image. For choosing the seed positions, simple strategies include random placement or positioning the seeds evenly over the image. A slightly more sophisticated approach is to put them at the brightest or darkest points in the image.

For real-time inspection tasks, there is often not enough time available to perform region growing. Thus, region growing is normally not applied in real-time inspection tasks. One particular application of region growing is in the inspection of pipes [Stefani et al., 1996]. It was found that region growing could give a more accurate segmentation of defects than thresholding or edge detection. The developed method consisted of a number of steps:

1.   Subdivide the image into non-overlapping squares with a dimension not larger than the diameter of the largest defect.

2.   Choose the seed pixel in each square to be the pixel with the lowest grey level.

3. Check the grey levels of neighbouring pixels and then add them to the region if their grey levels are close to those of the rest of the region.

4. Return to step 3 unless no more pixels can be added to the region or the squares are fully segmented.

5. If a region fills a sufficiently large percentage of a square then it is considered to be background. Also, a region is considered to be background if it fills a sufficiently small percentage of a square.

3.4 Split-and-merge

Split-and-merge is a top-down method of segmentation. First, the image is divided into several equal-sized regions. Second, each of these regions is analysed to determine if it is sufficiently homogeneous. If this is true then segmentation for that region terminates. Otherwise, the region is subdivided again. The process operates iteratively until acceptably uniform regions have been found. Then, adjacent regions are merged if they are found to be similar.

Figure 3.14 shows an image that has undergone three iterations of manual splitting. First, the image was divided into four equal-sized squares. Next, the squares were analysed for uniformity. All squares were split again into four, as they were not adequately homogenous. In the third iteration, only some of the squares are further divided, as the others are segmented satisfactorily.

A problem with the spilt-and-merge technique is to find the optimal similarity threshold. This threshold determines when to stop the splitting. If the threshold is set too low then splitting will continue until very small regions are found. Pixels that should belong to the same region would be placed into different regions. If the threshold is set too high then splitting will terminate early and very large inhomogeneous regions will be derived. Thus, areas that should be placed into separate regions would be put into the same region.

Like region growing, the split-and-merge technique is not normally applied in AVI applications. One example of where experiments have been carried out in using split-and-merge for inspection is the work of Ventura and Chen [1992]. The application was the inspection of curved industrial parts. A two-stage segmentation method was developed. First, the split-and-merge method was applied to approximate the object boundary in an iterative manner. Second, to find the best fitting polygon approximation, an end-point adjustment procedure was performed. The method was found to be accurate and fast.

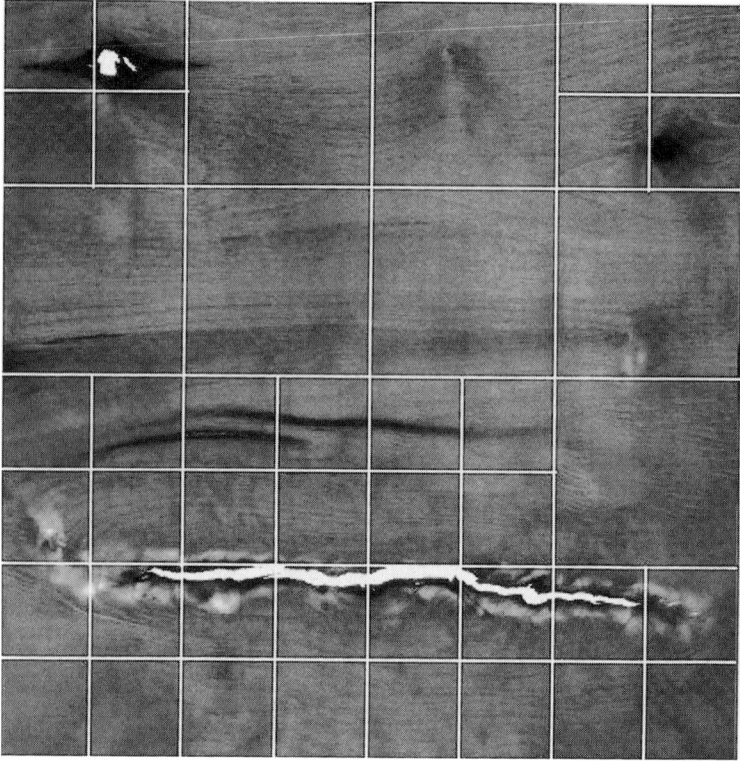

Figure 3.14 Split-and-merge segmentation after three levels of splitting

3.5 Window-based Subdivision

One method to avoid having to choose and develop a suitable image segmentation method is simply to divide the image into windows [Song et al., 1992]. Thus, each window can be analysed and its contents determined. This top-down technique has been employed in texture analysis, where it is required to determine what texture is in different parts of the image [Pham and Cetiner, 1996]. This will be explained in more detail in Chapter 6. Figure 3.15 shows an image of a textured surface from Broadatz [1968]. The white lines indicate how this image can be simply subdivided into equal-sized regions. These regions can then be classified to determine their contents.

When areas of interest occupy only a small percentage of the image, attempts have been made to write algorithms that can determine quickly which regions are potentially of interest. More time-consuming techniques can then be applied to those regions to determine their exact contents. This idea is particularly important for colour images because they require more computation. The algorithms suggested by Forrer et al. [1988] are called image sweep-and-mark (ISM) algorithms because they sweep through the image and, depending upon features calculated from each region, mark the regions which are likely to contain a defect. Forrer et al. [1989] tried three techniques called statistical image sweep-and-mark (SISM), morphological image sweep-and-mark (MISM) and colour-cluster image sweep-and-mark (CISM). It was found that SISM performed most favourably. Butler et al. [1989] later suggested neighbourhood statistical image sweep-and-mark (NSISM). This employed a lower threshold for regions whose neighbours had been marked as potentially defective. NSISM was found to perform better than SISM.

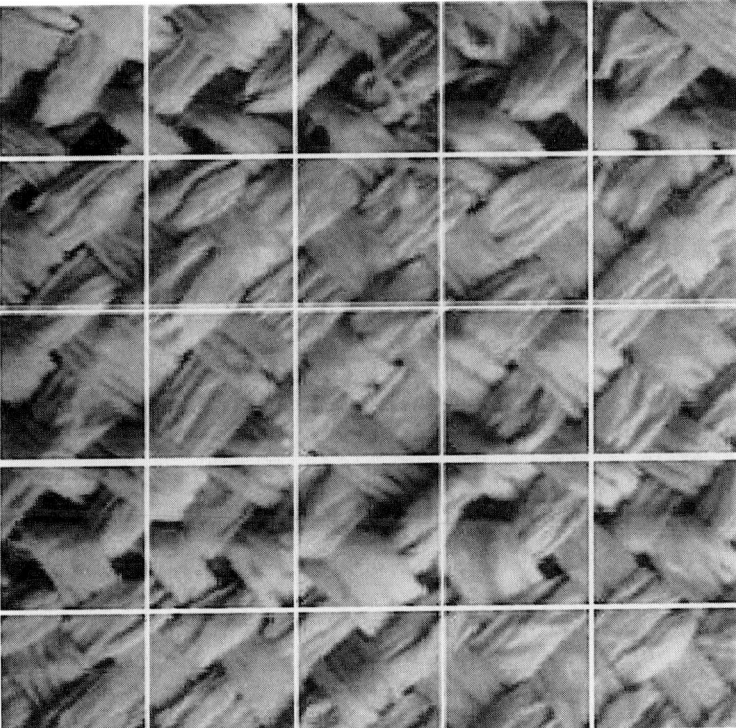

Figure 3.15 Simple subdivision of a textured image

Windows-based subdivision is fast but has disadvantages. First, a defect may be spread across two or more regions. Second, a small defect in a large region may not be found. Third, a region containing more than one defect cannot be classified. Another problem with the approach is that it is difficult to obtain size or shape information about a defect from the region. Classification is simplified given shape information because defects of the same type often have a similar shape and size.

3.6 Template Matching

Template matching is a top-down method of segmentation that is sometimes suitable for AVI tasks as it is normally known what type of object that will be in the image. In this method, a small template image is created, which contains the object that is to be searched for. This template is then moved across and down the main image and the correlation between the template and that point of the image is calculated. The calculation is based on the difference in grey level between the corresponding pixels in the template and in the image. If the match is significantly large then the object is found at that point of the image.

One of the major problems with template matching is that it requires a very large number of computations to be performed. There are several techniques that have been developed to increase the speed of template matching [Demant et al., 1999]. One technique is to use a coarse-to-fine search. The template is first moved over the image in large steps. If a suitably high correlation is found at a point, the template is moved over this area using smaller steps. Another technique is called sub-sampling, where just a sample of points from the template is used to calculate the match. The sub-sample can be spread evenly over the template or can be concentrated on more important parts of the template. It is also possible to increase computation time by stopping calculations early when the value exceeds a certain limit.

Further problems with template matching are that it is very sensitive to object rotations and image brightness adjustments.

Jo Coleman

Information Update Service

Butterworth-Heinemann

FREEPOST SCE 5435

Oxford

Oxon

OX2 8BR

UK

Keep up-to-date with the latest books in your field.

Visit our website and register now for our FREE e-mail update service, or join our mailing list and enter our monthly prize draw to win £100 worth of books. Just complete the form below and return it to us now! (FREEPOST if you are based in the UK)

www.bh.com

Please Complete In Block Capitals

Title of book you have purchased:..

....

Subject area of interest:..

Name:..

Job title:..

Business sector (if relevant):...

Street:..

Town:... County:...

Country:... Postcode:.......................................

Email:..

Telephone:...

How would you prefer to be contacted: Post ☐ e-mail ☐ Both ☐

Signature:... Date:...

☐ Please arrange for me to be kept informed of other books and information services on this and related subjects (✔ box if not required). This information is being collected on behalf of Reed Elsevier plc group and may be used to supply information about products by companies within the group.

FOR OFFICE USE ONLY

Butterworth-Heinemann,
a division of Reed Educational
& Professional Publishing Limited.
Registered office: 25 Victoria Street,
London SW1H 0EX.
Registered in England 3099304.
VAT number GB: 663 3472 30.

**BUTTERWORTH
HEINEMANN**

A member of the Reed Elsevier plc group

3.7 Horizontal and Vertical Profiling

Profiling is a simple and fast technique that can help in segmentation for some images. It operates by summing the pixel values in the horizontal or vertical directions. A dark line in an image causes a valley in the profile. Figure 3.16 shows an image of a wood board that contains a dark vertical defect called a streak. The vertical profile of the image is shown in Figure 3.17. It can be seen that the line in the image corresponds with the valley in the profile.

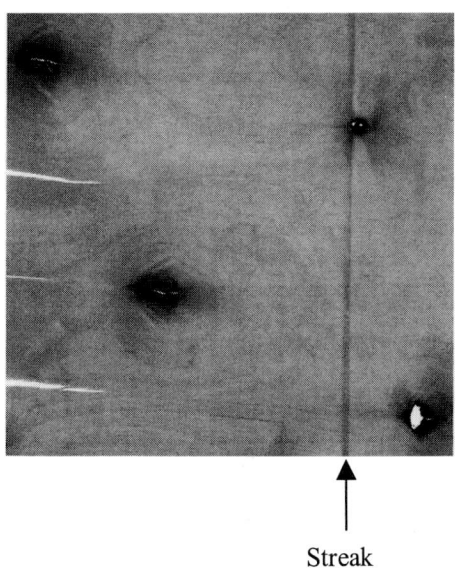

Streak

Figure 3.16 Image containing a dark vertical defect

3.8 AI-based Segmentation

Neural networks, fuzzy logic and intelligent optimisation techniques have been incorporated in segmentation by many researchers.

A method for overcoming the problem of designing a segmentation technique is to utilise a neural network to determine what each pixel in the image represents [Schmoldt et al., 1997]. Using a MLP for segmentation obviates the need to develop dedicated segmentation algorithms because it learns what constitutes an object without direct human intervention. The MLP can have nine inputs - the grey levels of the pixel to be classified and its immediate neighbours. The network will have as many outputs as there are different objects to be classified in the image. An advantage of this technique is that it can perform segmentation and classification in one single step. However, there are several problems. First, nine inputs to the network may not be sufficient. Second, the method is very time consuming without dedicated hardware. Third, large training sets are required.

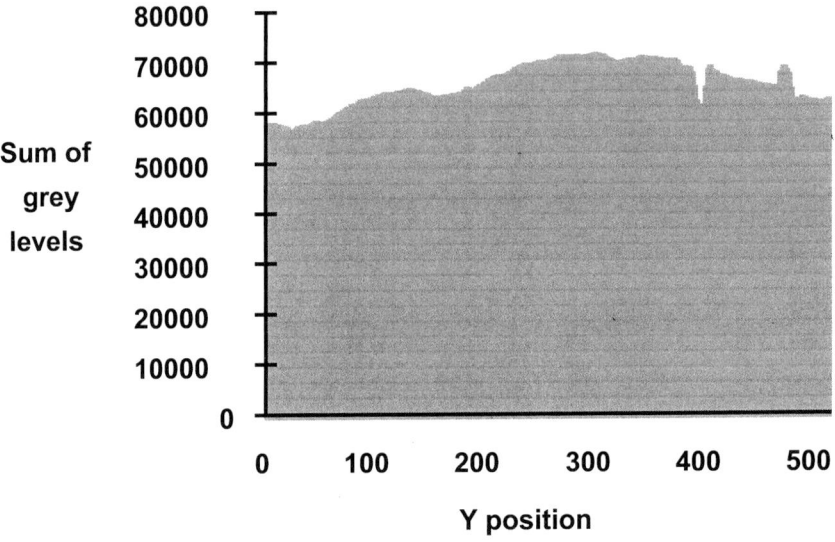

Figure 3.17 Vertical profile for image in Figure 3.16

Sziranyi and Csapodi [1998] performed image segmentation using cellular neural networks (CNNs). CNNs are ideal for real-time image analysis as they were designed for hardware realisation. The system was implemented using a programmable 22 x 20 CNN chip and the execution time was a few microseconds.

Fuzzy logic has been employed for image segmentation. Fang and Cheng [1992] worked on determining optimal thresholds for segmentation using fuzzy risk assessment. Other researchers attempted to perform segmentation using fuzzy clustering [Tolias and Panas, 1998]. Clustering involves grouping together items that have similar characteristics. It is relevant to segmentation because segmentation requires pixels with similar properties to be clustered into the same region.

Unsupervised neural networks are also utilised as clustering tools and so can also be applied in image segmentation.

Segmentation has also been performed with optimisation techniques. In the work of Bhandarkar and Zhang [1999], image segmentation is treated as a combinatorial optimisation problem. A cost function is derived based both on edge information and region grey-level homogeneity. A hybrid evolutionary strategy is employed, which is based on genetic algorithms and simulated annealing. This hybrid approach was found to be superior to the simple GA-based method.

3.9 Post-Processing of Segmented Images

After segmentation has been performed, two problems can be experienced. First, many areas, which are background, may be detected as objects. Second, an object may be split into several areas. This creates the need for a post-processing stage after initial segmentation. Two conventional methods for post-processing segmented images are morphology and the Hough transform. Other techniques, based on artificial intelligence, are also given in this section.

3.9.1 Morphology

Morphology, as described in the previous chapter, is a standard image-processing tool that can be used to process objects in segmented images. To remove small "salt and pepper type" regions from a segmented image, Kim and Koivo [1994] performed morphological opening. To join objects together, morphological closing can be employed. The advantage of morphology is that it is a simple operation and can be applied in real time using standard convolution filters.

3.9.2 Hough Transform

The Hough Transform is a well-established image processing technique [Gonzalez and Woods, 1992]. It can be utilised for finding pixels in an image that are part of the same straight line. For each pixel at point (x, y) in the image, values are generated which satisfy the equation of a straight line $y = mx + c$. For each m and c generated, position (m, c) in the Hough space is incremented. Then, large values are searched for in the Hough space because they correspond to straight lines in the image. It is then possible to join pixels that are close to each other and fall along the same straight line. The Hough Transform can also be adapted to detect other shapes such as circles. The main problem with applying the Hough Transform for processing segmented images is that it is very time consuming and so difficult to implement in real-time applications. This is because it is necessary to generate a large number of values of m and c for every pixel.

3.9.3 AI-based Post Processing

Pham and Alcock [1998] proposed two new techniques for post-processing segmented images. The techniques were inspired by the artificial intelligence techniques of fuzzy logic and self-organising neural networks.

For segmentation, it was proposed to use intermediate grey levels, so pixels that were considered more likely to be objects were given a higher grey level than those resembling background. Then, pixels touching each other were formed into objects and the *evidence* of each object was determined by summing its grey levels. Any object with evidence below a specified value was then removed.

To join together objects representing the same real-world entity, a self-organising neural network technique inspired by ART-1 was developed. The objects were sorted into order of size and then were presented to the network in that order. If new objects were considered to be part of an existing neuron, based on a distance function, then the new object was incorporated into the neuron. Otherwise, a new neuron was created to store the object. Finally, all objects stored in each neuron were grouped into the same region. Further details of these two techniques are found in Chapter 6.

3.10 Discussion

In industrial AVI tasks, products arrive for inspection in rapid succession. For each product, typically there is around one second or even less to perform the complete inspection. This involves the time not just for segmentation but also for image acquisition, enhancement, feature extraction and classification. Segmentation can be a time-consuming process and so a larger proportion of the total inspection time than for the other stages can be assigned to it. However, this still means that well under one second is available. Therefore, segmentation techniques should take advantage as much as possible of the hardware-based image processing functions of the vision system.

To simplify the segmentation requirements, a common approach is to design the lighting system carefully so that simple and fast global thresholding can be employed. If post-processing of objects in the segmented image is required, morphological operations can be carried out, as these will operate quickly. However, in the future, as inspection hardware increases in speed, more AI techniques will be able to be incorporated into industrial AVI systems.

3.11 Summary

Segmentation of images can be a complex task. The most popular techniques for segmentation are edge detection, thresholding and region growing.   Artificial intelligence techniques have also been applied for segmentation and have given good performance. However, all techniques suffer from drawbacks. Currently, the main drawback with AI techniques is that they are difficult to implement efficiently with inspection hardware.

No segmentation technique is perfect for all applications and so segmentation must be designed carefully together with the image acquisition stage.   Due to the strict time-constraints imposed in AVI applications, simple segmentation methods are recommended. These often work well when combined with effective illumination.

References

Abdou I.E. and Pratt W.K. (1979) Quantitative Design and Evaluation of Enhancement/Thresholding Edge Detectors. *Proc. IEEE*. Vol. 67, No. 5, pp. 753 - 763.

Acton S.T. and Mukherjee D.P. (2000) Area Operators for Edge Detection. *Pattern Recognition Letters*. Vol. 21, No. 8, pp. 771 - 777.

Batchelor B.G. and Whelan P.F. (1997) *Intelligent Vision Systems for Industry*. Springer-Verlag, Berlin and London.

Bhandarkar S.M. and Zhang H. (1999) Image Segmentation using Evolutionary Computation. *IEEE Trans. on Evolutionary Computation*. Vol. 3, No. 1, pp. 1 - 21.

Broadatz P. (1968) *Textures: A Photographic Album for Artists and Designers*. Van Nostrand Reinhold, New York.

Butler D.A., Brunner C.C. and Funck. J.W. (1989) A Dual-threshold Image Sweep-and-Mark Algorithm for Defect Detection in Veneer. *Forest Products Journal.* Vol. 39, No. 5, pp. 25 - 28.

Das D.N. and Mukhopadhyay S. (1998) Image Edge Detection and Enhancement by an Inversion Operation. *Applied Optics.* Vol. 37, No. 35, pp. 8254 - 8257.

De Santis A. and Sinisgalli C. (1999) A Bayesian Approach to Edge Detection in Noisy Images. *IEEE Trans. on Circuits and Systems. Part 1 - Fundamental Theory and Applications.* Vol. 46, No. 6, pp. 686 - 699.

Demant C., Streicher-Abel B. and Waszkewitz P. (1999) *Industrial Image Processing: Visual Quality Control in Manufacturing.* Springer-Verlag, Berlin.

Fang N. and Cheng M.C. (1992) On Threshold Selection using Fuzzy Risk Criterion. *Japanese Journal of Applied Physics. Part 1 - Regular Papers, Short Notes and Review Papers.* Vol. 31, No. 5A, pp. 1382 - 1388.

Forrer J.B., Butler D.A., Funck J.W. and Brunner C.C. (1988) Image Sweep-and-Mark Algorithms, Part 1, Basic Algorithms. *Forest Products Journal.* Vol. 38, No. 11/12, pp. 75 - 79.

Forrer J.B., Butler D.A., Brunner C.C. and Funck J.W. (1989) Image Sweep-and-Mark Algorithms, Part 2, Performance Evaluations. *Forest Products Journal.* Vol. 39, No. 1, pp. 39 - 42.

Gonzalez R.C. and Woods R.E. (1992) *Digital Image Processing* (3rd ed.). Addison-Wesley, Reading, MA.

Gudmundsson M., El Kwae E. and Kabuka M.R. (1998) Edge Detection in Medical Images using a Genetic Algorithm. *IEEE Trans. on Medical Imaging.* Vol. 17, No. 3, pp. 469 - 474.

Heinemann P.H., Hughes R., Morrow C.T., Sommer H.J., Beelman R.B. and West P.J. (1994) Grading of Mushrooms using a Machine Vision System. *Trans. of the ASAE.* Vol. 37, No. 5, pp. 1671 - 1677.

Hou T.H. and Kuo W.L. (1997) A New Edge Detection Method for Automatic Visual Inspection. *Int. Journal of Advanced Manufacturing Technology.* Vol. 15, No. 7, pp. 713 - 721.

Kim C.W. and Koivo A.J. (1994) Hierarchical Classification of Surface Defects on Dusty Wood Boards. *Pattern Recognition Letters.* Vol. 15, No. 7, pp. 713 - 721.

Kothari R., Klinkhachorn P. and Huber H.A. (1991) A Neural Network Based Histogramic Procedure for Fast Image Segmentation. *Proc. 23rd Sym. on System Theory*. Columbia, SC, pp. 203 - 207.

Murtovaara S., Juuso E. and Sutinen R. (1996) Fuzzy Logic Edge Detection Algorithm. *Proc. 3rd Int. Workshop on Image and Signal Processing*. Manchester, UK. pp. 423 - 426.

Otsu N. (1979) A Threshold Selection Method from Grey-Level Histograms. *IEEE Trans. on Systems, Man and Cybernetics*. Vol. 9, No. 1, pp. 62 - 66.

Pavlidis T. (1992) Why Progress in Machine Vision is so Slow. *Pattern Recognition Letters*. Vol. 13, No. 4, pp. 221 - 225.

Pham D.T. and Alcock R.J. (1996) Automatic Detection of Defects on Birch Wood Boards. *Proc. IMechE. Part E - Journal of Process Mechanical Engineering*. Vol. 210, pp. 45 - 52.

Pham D.T. and Alcock R.J. (1998) Artificial Intelligence Techniques for Processing Segmented Images of Wood Boards. *Proc. IMechE. Part E - Journal of Process Mechanical Engineering*. Vol. 212, pp. 119 - 129.

Pham D.T. and Bayro-Corrochano E.J. (1992) Neural Computing for Noise Filtering, Edge Detection and Signature Extraction. *Journal of Systems Engineering*. No. 2, pp. 111 - 122.

Pham D.T. and Cetiner B.G. (1996) A New Method for Describing Texture. *Proc. 3rd Int. Workshop on Image and Signal Processing*. Manchester, UK, pp. 187 - 190.

Russo F. (1998) Edge Detection in Noisy Images using Fuzzy Reasoning. *IEEE Trans. on Instrumentation and Measurement*. Vol. 47, No. 5, pp. 1102 - 1105.

Sahoo P.K., Soltani S., Wong A.K.C. and Chen Y.C. (1988) A Survey of Thresholding Techniques. *Computer Vision, Graphics and Image Processing*. Vol. 41, pp. 233 - 260.

Sanby C., Norton-Wayne L. and Harwood R. (1995) The Automated Inspection of Lace using Machine Vision. *Mechatronics*. Vol. 5, No. 2-3, pp. 215 - 231.

Schmoldt D.L., Li P. and Abbott A.L. (1997) Machine Vision Using Artificial Neural Networks with Local 3D Neighborhoods. *Computers and Electronics in Agriculture*. Vol. 16, No. 3, pp. 255 - 271.

Song K.Y., Petrou M. and Kittler J. (1992) Texture Defect Detection: A Review. *SPIE Vol. 1708: Applications of Artificial Intelligence X: Machine Vision and Robotics*. Orlando, FL., pp. 99 - 106.

Stefani S.A., Nagarajah C.R. and Toncich D. (1996) Non-Contact Inspection for the Detection of Internal Surface Defects in Hollow Cylindrical Work-Pieces. *Int. Journal of Advanced Manufacturing Technology*. Vol. 11, pp. 146 - 154.

Sundaram R. (1999) Algorithms for Adaptive Transform Edge Detection. *IEEE Trans. on Signal Processing*. Vol. 47, No. 8, pp. 2313 - 2317.

Sziranyi T. and Csapodi M. (1998) Texture Classification and Segmentation by Cellular Neural Networks using Genetic Learning. *Computer Vision and Image Understanding*. Vol. 71, No. 3, pp. 255 - 270.

Tolias Y.A. and Panas S.M. (1998) Image Segmentation by a Fuzzy Clustering Algorithm using Adaptive Spatially Constrained Membership Functions. *IEEE Trans. on Systems, Man and Cybernetics. Part A - Systems and Humans*. Vol. 28, No. 3, pp. 359 - 369.

Truchetet F. and Cholley J.P. (1997) Tampoprint Quality Control by Artificial Vision. *Materials Evaluation*. Vol. 55, No. 12, pp. 1361 - 1366.

Ventura J.A. and Chen J.M. (1992) Segmentation of 2-Dimensional Curve Contours. *Pattern Recognition*. Vol. 25, No. 10, pp. 1129 - 1140.

Wen W. and Xia A. (1999) Verifying Edges for Visual Inspection Purposes. *Pattern Recognition Letters*. Vol. 20, No. 3, pp. 315 - 328.

Zhou L.Y., Chalana V. and Kim Y. (1998) PC-based Machine Vision System for Real-time Computer-aided Potato Inspection. *Int. Journal of Imaging Systems and Technology*. Vol. 9, No. 6, pp. 423 - 433.

Zhu S.Y., Plataniotis K.N. and Venetsanopoulos A.N. (1999) Comprehensive Analysis of Edge Detection in Color Image Processing. *Optical Engineering*. Vol. 38, No. 4, pp. 612 - 625.

Problems

1. Describe the advantages and disadvantages of thresholding, windows-based subdivision and template matching for segmentation.

2. Show the results of using thresholds of 50, 80 and 100 on the following image:

82	107	103	112
110	43	67	87
113	56	74	119
120	105	112	101

3. Show the effect of the Robert's edge detector on the image of Problem 2.

4. Using the image of Problem 2 and starting at the pixel with a grey level of 43, show the effect of region growing with thresholds of 10, 30 and 50.

5. Calculate the vertical profile of the following image region:

35	27	19	29	37	12	38
30	25	23	29	32	14	32
35	28	18	22	31	14	30

Describe the differences between the valleys found.

6. Discuss reasons why machine segmentation of images is far behind the capabilities of human vision.

7. In ProVision, the thresholding function is called *W Binarization*. Open the project DATE, which incorporates this function. Open the *W Binarization* dialog box and view the effect of thresholding by clicking the *Test* and *Display* buttons.

8. In the DATE project, describe the effect of changing the upper threshold to the following values: 20, 120 and 220.

Chapter 4

Feature Extraction and Selection

The result of segmentation is an image containing one or more objects. From each of these objects, a set of features is extracted to describe the object. This set of features is called a feature vector, which is presented to a classifier to identify the type of the object. It is important that a suitable feature vector is chosen, as this is the only data passed to the classifier. Many different features have been, and continue to be, developed. The features described in this chapter are ones that have been applied frequently in AVI tasks.

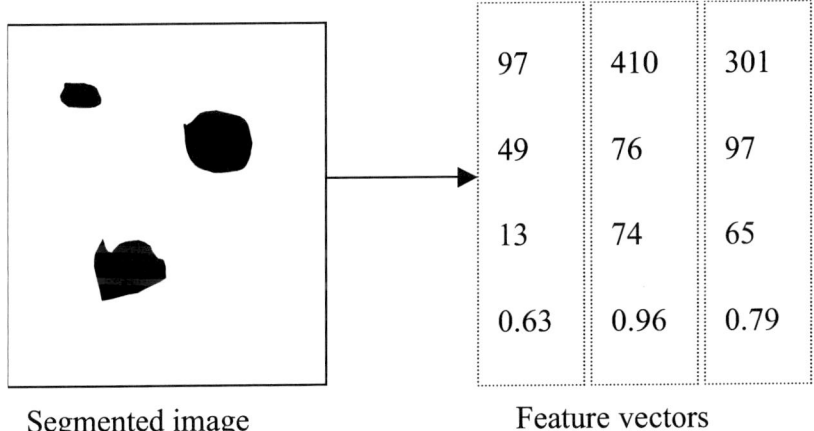

Segmented image   Feature vectors

Figure 4.1 Feature extraction process

Figure 4.1 illustrates the feature extraction process. On the left is a segmented image containing three objects. For each object, a feature vector is derived, consisting of four measurements. The feature extraction stage is different from the previous stages of the AVI process because its output is normally in the form of a set of numbers rather than an image.

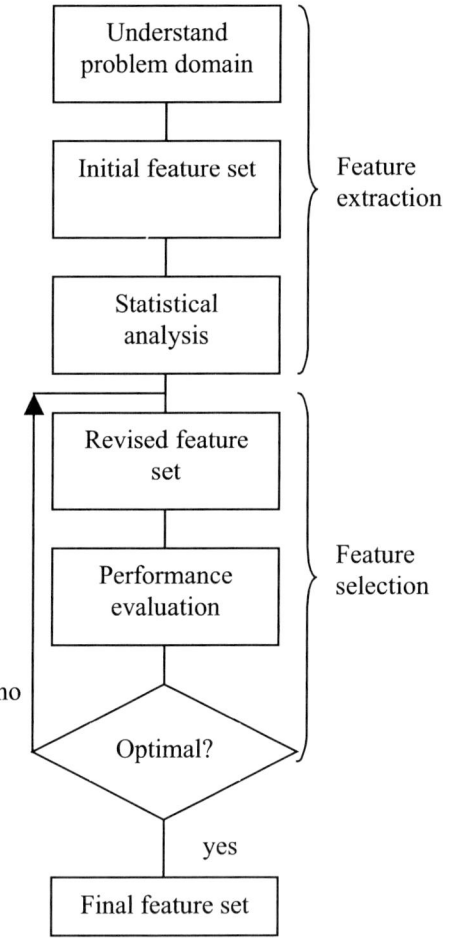

Figure 4.2 Feature set development

Ideally, the development of a suitable feature vector should consist of a number of stages. The early stages of feature development are very important but are often overlooked. Figure 4.2 illustrates the steps of feature development [Webb, 1999]. First, it is important to understand the problem domain. This involves determining the measurements that characterise the object effectively. Once an initial set of features has been found, data sets can be generated. Next, it is informative to perform statistical analysis on the extracted data sets. This can take the form of summary statistics. Figure 4.3 shows a summary table for a sample of the benchmark IRIS flower classification problem [Blake et al., 2002]. Another important method for manual analysis of features is graphical plots. These are useful, as humans are better able to visualise graphics than numbers. However, it is only physically possible to plot three features at one time. For clear viewing on paper or a computer screen, normally only two features are plotted. Thus, if there is a large number of features, there are many combinations of features to generate plots for. Figure 4.4 shows the plot of two features from the IRIS dataset. It can be seen that one of the classes is clearly separable from the other two using just one feature (Petal length).

	Feature 1			Feature 2		
	Mean	Standard deviation	Range	Mean	Standard deviation	Range
Overall	5.4	0.60	4.3 - 6.7	3.1	0.42	2.3 - 4.4
Class 1	5.0	0.37	4.3 - 5.7	3.5	0.44	3.0 - 4.4
Class 2	5.8	0.51	4.9 - 6.7	2.8	0.29	2.3 - 3.2

Figure 4.3 Summary statistics for a sample of IRIS data

After a set of features has been developed and analysed, the next stage is to evaluate the features and choose the best subset. This stage is normally iterative and is called feature selection.

Features can be extracted either from *windows* or directly from segmented *objects* [Song et al., 1992]. Naturally, features extracted from windows are called window features and those from objects are object features. Mainly, features characterise tonal and shape properties. The features explained here are given for grey-scale images. For colour images, these features can be calculated for each of the three colour channels.

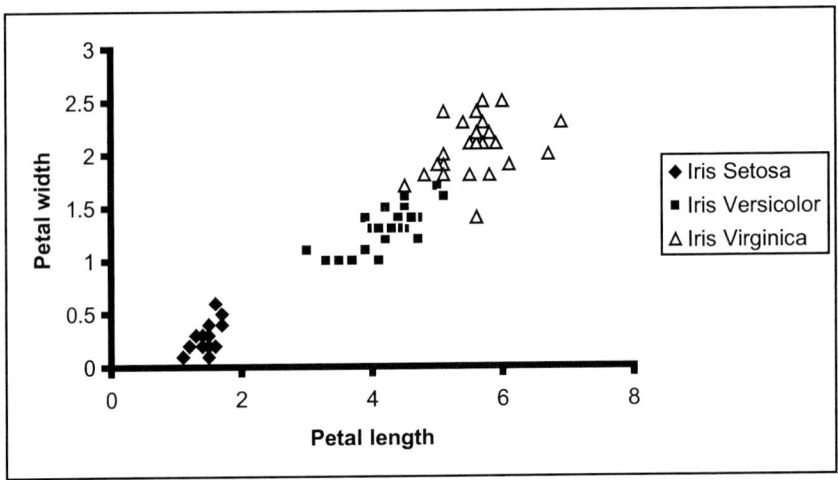

Figure 4.4 Feature plot

When several different features are used together, it is probable that each feature will have a different range. To bring the features into the working range of most classifiers and to reduce bias during classification, it is a normal practice to scale the features between 0 and 1. A common formula for achieving this is:

$$X_s = \frac{X_u - X_{min}}{X_{max} - X_{min}} \tag{4.1}$$

where X_s is the scaled value of the feature, X_u is the unscaled value of the feature, X_{min} is the minimum value that the feature takes and X_{max} is its maximum value. X_{min} and X_{max} are estimated from the training data set. If with new data, X_s has a value of less than zero or greater than one, hard limiting can be employed. Data can be scaled between ± 1 with the following formula:

$$X_s = \left(2 * \left(\frac{X_u - X_{min}}{X_{max} - X_{min}} \right) \right) - 1 \tag{4.2}$$

It is also possible to scale using the mean and standard deviation of the data. This can be calculated as:

$$X_s = \frac{X_u - X_\mu}{3X_\sigma} \tag{4.3}$$

where X_μ is the mean value of X_u for the training data and X_σ is its standard deviation. In practical applications, the type of scaling used should have little real effect on the classification results.

4.1 Window Features

Window features are extracted after the image has been subdivided into equal-sized regions. For windows, tonal features are called first-order features. Features that are designed to determine the shape of objects contained in the window are second-order features. Two of the most common second-order features are Fourier transform features and co-occurrence features.

4.1.1 First-Order Features

Typical first-order features are the statistical features of mean (μ), standard deviation (σ), skewness and kurtosis. The advantage of using statistical features is that they have a theoretical basis. To calculate these features, first the grey-level histogram of the window is calculated. This is a one-dimensional array, where $h(x)$ represents the number of pixels in the window with a grey level of x. x can take any value between 0 and $Z-1$, where Z is the number of grey levels in the image.

$$\mu = \frac{\sum_{x=0}^{Z-1} h(x)}{Z} \tag{4.4}$$

$$\sigma = \sqrt{\frac{\sum_{x=0}^{Z-1} (h(x) - \mu)^2}{Z}} \tag{4.5}$$

$$Skewness = \frac{\sum_{x=0}^{Z-1}(h(x) - \mu)^3}{Z\sigma^3} \tag{4.6}$$

Skewness shows by how much the distribution of grey levels deviates from that of a normal distribution. If the distribution is skewed to the right, then the skewness factor is positive. The factor is negative if the distribution is skewed to the left.

$$Kurtosis = \frac{\sum_{x=0}^{Z-1}(h(x) - \mu)^4}{Z\sigma^4} - 3 \tag{4.7}$$

Kurtosis shows how "peaked" the distribution is compared to a normal distribution. Kurtosis is positive if the peak in the histogram is higher than that of a normal distribution. If the peak is lower, kurtosis is negative. For colour images, these features can be calculated for each of the three colour components (red, green and blue), giving twelve first-order features.

Sobey and Semple [1989] found that the mean is a very important feature and that the best results are obtained when it is used with kurtosis and standard deviation. Skewness was found to be of marginal benefit.

Apart from the above features, there are others that could be employed, such as dispersion, entropy and energy.

$$Dispersion = \sum_{x=0}^{Z-1}|h(x) - \mu| \tag{4.8}$$

$$Entropy = \sum_{x=0}^{Z-1}h(x).\log_2(h(x)) \tag{4.9}$$

$$Energy = \sum_{x=0}^{Z-1}x^2.h(x) \tag{4.10}$$

It is possible to divide equations (4.8) - (4.10) by the number of grey levels in the images, Z. However, for image classification purposes, this would not have any effect on the final classification result.

Other common first-order features that could be employed include:

- lower quartile (a):

$$a = y \mid \sum_{x=0}^{x=y} h(x) \geq 0.25s \; and \; \sum_{x=0}^{x=y-1} h(x) < 0.25s$$

(4.11)

where:

$$s = \sum_{x=0}^{x=Z-1} h(x)$$

- upper quartile (b):

$$b = y \mid \sum_{x=0}^{x=y} h(x) \geq 0.75s \; and \; \sum_{x=0}^{x=y-1} h(x) < 0.75s$$

(4.12)

- lowest grey level (c):

$$c = x \mid h(x) \neq 0$$

where $0 \leq x < Z$ and $h(i) = 0 \; \forall i: 0 \leq i < x$

(4.13)

- highest grey level (d):

$$d = x \mid h(x) \neq 0$$

where $0 \leq x < Z$ and $h(i) = 0 \; \forall i: x < i < Z$

(4.14)

- histogram tail length on dark side (e):

$$e = a - c$$

(4.15)

- histogram tail length on light side (f):

$$f = d - b$$

(4.16)

- median grey level (g):

$$g = o\left(\frac{s}{2}\right)$$

$$(4.17)$$

where o(j) is the grey level of the jth pixel when the pixel grey levels in the window are ordered.

- range of grey levels (h):

$$h = d - c$$

$$(4.18)$$

- inter-quartile range (i):

$$i = b - a$$

$$(4.19)$$

- mode grey level (j):

$$j = x \mid h(x) > h(i) : \forall i, 0 \leq i, x < Z, i \neq x$$

$$(4.20)$$

4.1.2 Fourier Features

A common set of second-order features is derived from the Fourier Transform (FT) [Russ, 1995]. The FT converts an image from the spatial domain into the frequency domain. It was originally designed for continuous signals and has been often employed in electronics signal processing. The basis of the FT is to approximate a signal as a series of sine and cosine waves of increasing frequency. For discrete signals, such as images, the Discrete Fourier Transform (DFT) is used. In real-time analysis, it is important to implement the FT efficiently so that it does not have a long execution time. A powerful algorithm for this is the Fast Fourier Transform (FFT).

The FT has a mathematical basis and requires integrating a function involving complex numbers. For a non-mathematician, it is not obvious how the equations should be implemented for image processing. For interested readers, the mathematics underlying the FT can be found in [Russ, 1995]. To facilitate understanding, Russ also gave an algorithm, found in Figure 4.5, to explain how the FT could be realised for image processing.

```
Complex_numbers F(512), U, W, T, CMPLX
PI = 3.1415927
N = 2^LN
NV2 = N/2
NM1 = N-1
J=1
for (I=1 to NM1)           /* Input re-ordering */
{
        if (I < J)
        {
                T = F(J)
                F(J) = F(I)
                F(I) = T
        }
        K = NV2
        while (K < J)
        {
                J = J-K
                K=K/2
        }
        J=J+K
}
for (L=1 to LN)            /* Frequency doubling */
{
        LE = 2^L
        LE1 = LE/2
        U = (1.0, 0.0)
        W = CMPLX(Cos(PI/LE1), -Sin(PI/LE1))
        for (J=1 to LE1)
        {
                for (I=J to N, step size=LE)
                {
                        IP = I + LE1
                        T = F(IP) * U
                        F(IP) = F(I) - T
                        F(I) = F(I) + T
                }
                U=U*W
        }
}
for (I=1 to MN)           /* Output normalising */
        F(I) = F(I) / FLOAT(N)
```

Figure 4.5 Fourier transform algorithm [Russ, 1995]

The FT is one-dimensional and can be applied to each row of the image in turn. This is not the optimal method for implementing the FFT but is simple and frequently used. The inputs to Russ's algorithm are a one-dimensional array F, which is the row of the image being analysed, and LN, the number of frequencies to be used in the output. The algorithm contains three loops. The first reorders the input, the second performs the frequency doubling required by the FT and the third normalises the output.

The algorithm produces a complex number as the result for each pixel. As complex values cannot be displayed in an image, normally just the magnitude is used and not the phase information.

4.1.3 Co-occurrence Features

The most widely-used second-order features have been *co-occurrence* features [Haralick et al., 1973]. These depend upon the relationships of the grey levels of pixels that are close to each other. The co-occurrence matrix is calculated and then several different features can be derived from this matrix.

To calculate co-occurrence features, first, the data is quantised into Q levels. Figure 4.6 shows an example of grey levels from a window of size 4x4 in an image with 256 grey levels. The result of quantising this image into 4 levels (Q=4) is shown in Figure 4.7.

135	130	100	55
132	111	60	57
140	137	93	51
150	142	101	52

Figure 4.6 Grey levels in a window of size 4x4

2	2	1	0
2	1	0	0
2	2	1	0
2	2	1	0

Figure 4.7 Result of quantisation on the window

Next, a two dimensional matrix $c(i, j)$ is constructed ($0 < i,j < Q$). Point (i, j) in the matrix represents the number of times that a pixel in the sequence with level i is followed, at a distance d and angle θ, by a pixel with level j. A number of these matrices can be calculated for angles at intervals of 45° and for each value of d. Figure 4.8 shows the matrix created for the window with d equal to 1 and θ equal to 0°.

	0	1	2	3
0	1	4	0	0
1	4	0	4	0
2	0	4	3	0
3	0	0	0	0

Figure 4.8 Co-occurrence matrix for the window

After the matrices have been determined, co-occurrence features are calculated. A large number of features can be derived from the matrices. Here, seven common co-occurrence features are given. These are energy, entropy, correlation (COR), inertia, local homogeneity (LH), cluster shade (CS) and cluster prominence (CP).

$$Energy = \sum_{i=1}^{Q} \sum_{j=1}^{Q} c(i, j)^2$$

$$(4.21)$$

$$Entropy = \sum_{i=1}^{Q} \sum_{j=1}^{Q} c(i, j) . \log(c(i, j))$$

$$(4.22)$$

$$COR = \frac{\sum_{i=1}^{Q} \sum_{j=1}^{Q} (i - \mu_x)(j - \mu_y) . c(i, j)}{\sigma_x \sigma_y}$$

$$(4.23)$$

where:

$$\mu_x = \frac{\sum_{i=1}^{Q} i \sum_{j=1}^{Q} c(i, j)}{Q}$$

$$(4.24)$$

$$\mu_y = \frac{\sum\limits_{j=1}^{Q} j \sum\limits_{i=1}^{Q} c(i,j)}{Q}$$

(4.25)

$$\sigma_x^2 = \frac{\sum\limits_{i=1}^{Q} (i - \mu_x)^2 \sum\limits_{j=1}^{Q} c(i,j)}{Q}$$

(4.26)

$$\sigma_y^2 = \frac{\sum\limits_{j=1}^{Q} (j - \mu_y)^2 \sum\limits_{i=1}^{Q} c(i,j)}{Q}$$

(4.27)

$$Inertia = \sum_{i=1}^{Q} \sum_{j=1}^{Q} (i - j)^2 c(i,j)$$

(4.28)

$$LH = \sum_{i=1}^{Q} \sum_{j=1}^{Q} \frac{1}{1 + (i - j)^2} c(i,j)$$

(4.29)

$$CS = \sum_{i=1}^{Q} \sum_{j=1}^{Q} (i + j + \mu_x + \mu_y)^3 c(i,j)$$

(4.30)

$$CP = \sum_{i=1}^{Q} \sum_{j=1}^{Q} (i + j + \mu_x + \mu_y)^4 c(i, j)$$

(4.31)

Conners et al. [1983] found that co-occurrence features give a better performance than first-order features but that the best results are obtained when they are combined. To save valuable inspection time, they proposed that areas of interest could be found using fast first-order features. More detailed processing could then be performed on these areas of interest with co-occurrence features. Research by Sobey [1990] found that the best co-occurrence features to choose, when combined with tonal measures, were *cluster prominence* and *cluster shade*.

Co-occurrence features have been compared with other second-order features both theoretically [Conners and Harlow, 1980] and empirically [Weszka et al., 1976]. The features used in both studies were Fourier power spectrum, co-occurrence features, grey-level difference statistics and run-length statistics. Conners et al. found that co-occurrence features were the best whilst Weszka et al. discovered that all were comparable except the Fourier features, which performed worse than the others.

A major problem with co-occurrence features is that they are very computationally intensive and so are not applicable to real-time inspection. For this reason, new second-order window features have been developed to give the same information but with a much lower computational burden. One example of such features, developed by Pham and Cetiner [1996], will be given in Chapter 6.

4.2 Object Features

If the image is segmented and objects are found, features are extracted from these objects. This section gives commonly-employed object features.

4.2.1 Object Shape Features

The shape features given here are called X_length, Y_length, area, elongation, circularity, perimeter and radius. These are examples of the large number of shape features that could be utilised. X_length and Y_length are calculated using the co-ordinates (start_x, end_x, start_y, end_y) of the Minimum Bounding Rectangle (MBR) of the object. The MBR of an object is the smallest rectangle that can completely enclose the object and is also parallel to the edges of the image. An example of a MBR is shown in Figure 4.9.

$$X\_length = end\_x - start\_x \qquad (4.32)$$

$$Y\_length = end\_y - start\_y \qquad (4.33)$$

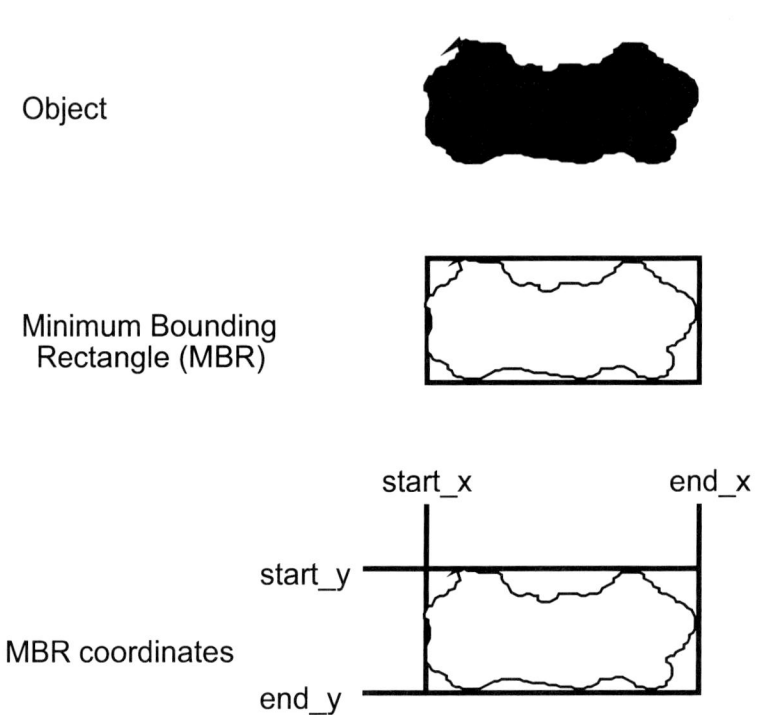

Object

Minimum Bounding
Rectangle (MBR)

MBR coordinates

Figure 4.9 Example of a minimum bounding rectangle

$$Elongation = \frac{X\_length}{Y\_length} \qquad (4.34)$$

Elongation is the ratio of the width of the MBR to its height. It is a measure of whether the object is horizontally or vertically elongated. The feature will give less meaningful values for objects that are diagonal lines in the MBR.

To calculate the perimeter of an object, it is first necessary to find the object's edge pixels. Next, the edge pixels must be traced. The distance between two pixels that can be connected by a horizontal or vertical line is 1. If the two pixels need to be connected by a diagonal line then the distance between them is $\sqrt{2}$. A less accurate but faster method of estimating the perimeter of an object is to calculate the number of edge pixels it contains.

Applying the Laplacian filter to a binary image and then counting the number of white pixels can perform this operation. The Laplacian filter can be executed quickly using a standard convolution filter (Figure 4.10).

0	-1	0
-1	4	-1
0	-1	0

Figure 4.10 Laplacian convolution filter

$$Circularity = \frac{4\pi.area}{(perimeter)^2} \tag{4.35}$$

If an object is a circle then its circularity value will be one. Less circular objects have smaller values and objects that are highly elongated have values of circularity close to zero.

The area of the object is often a useful feature for classification. In very simple classification tasks, the size of an object can give a good initial indication of what the object represents. This measure can be calculated simply by finding the number of pixels contained in the object.

The radius of an object is the distance from its centre to the edge. Two main measures can be calculated here, the minimum radius Rmin and the maximum radius Rmax. From these, the ratio between the two can be calculated: Rmax/Rmin. A further possibility is to plot the sequence of radii values as a histogram, from which features are extracted. The raw radii values can also be used directly as features themselves.

Other potentially useful features for classifying an object are the number of holes that it contains and its Euler number. The Euler number of an object is the number of connected components minus the number of holes. In most cases, the number of connected components is one.

4.2.2 Object Shade Features

As well as features that describe the shape of each object, it is useful to have features that characterise their shade or tone. For this, the histogram of the pixel values within the object is calculated. Then, various features are computed, such as the statistical features given in equations (4.4) - (4.7).

Pham and Alcock [1999] used linguistic variables as the basis for new features. Pixels were divided into five categories and the proportion of pixels in each category was calculated. The categories were called very dark, dark, mid-range, bright and very bright. Tests found that these performed better than the traditional statistical features.

Then, experiments were carried out on seventy-five images of wood boards to determine whether window features or object features give a better performance. Table 4.1 shows the results. It was found that each set of features gave approximately the same performance. However, when both were combined together, a performance improvement of around 10% was achieved. This highlights the need to have sufficient features.

	Run 1	Run 2	Run 3	Average
Object features	76.1%	69.3%	75.0%	73.5%
Window features	73.9%	83.0%	69.3%	75.7%
Combined features	86.4%	81.8%	87.5%	85.2%

Table 4.1 MLP classification accuracies with object and window features

4.3 Features from Colour Images

When colour images are employed, the number of possible features increases by more than three times. Colour images are similar to three grey-scale images, where images R, G and B show the intensities of the red, green and blue components, respectively. These images can be combined in different ways before feature extraction. The simplest combination is to find the mean of the three components ((R+G+B)/3). The effect of this is the same as if a simple grey-scale image were used. Alternatively, images can be treated separately and features extracted from each in turn. This gives three times as many features as would be derived from a grey-scale image. To measure the excess red in the image, the following image combination could be employed: 2R-G-B. Likewise, the excess image of the other two components can be generated.

Instead of extracting features from the colour images, the three components can be converted into a different colour space. The most common transformation is to the HSI space, with the components hue (H), saturation (S) and intensity (I). The equations for transforming from the RGB space to the HSI space, as given by Gonzalez and Woods [1992], are:

$$I = \frac{R+G+B}{3} \tag{4.36}$$

$$S = 1 - \frac{3a}{R+G+B} \tag{4.37}$$

where a is the minimum of R, G and B.

$$H = \frac{\cos^{-1}(0.5 + ((R-G)+(R-B)))}{\sqrt{((R-G)^2 + (R-B)*(G-B))}} \tag{4.38}$$

The value of H is meaningless if S is zero. H is an angle and, as such, takes a value between 0 and 360.

In human vision, the combination of hue and saturation together is important for recognition [Gonzalez and Woods, 1992]. Thus, in transferring tasks, such as the inspection of coloured fruits, from humans to machines, it can be beneficial to employ the HSI transform. For inspecting wood, Brunner et al. [1992] investigated converting the usual RGB colour space into other potentially more useful colour spaces. Five colour transforms, including HSI, were tested but it was found that none gave any improvement over RGB.

4.4 Feature Selection

Much research in AVI has been performed into improving segmentation and classification techniques. However, relatively little has been carried out on determining the best feature vector. Selection of the best features is one of the key factors in improving classifier performance. It is important to have sufficient features. However, features that do not improve classification accuracy should not be included.

The problem of feature selection is to generate a large number of features and then to select the subset of features that gives the best classification performance. Thus, it is an optimisation problem. There is no widely-recommended technique for feature selection. However, Castleman [1979] outlined some important factors that should be taken into account when selecting features:

1. For objects of different types, the features should take significantly different values. In other words, they should have a high inter-class variation.

2. For objects of the same type, the features should take similar values. That is, they should have a low intra-class variation.

3. The features should be independent. Each feature should represent a different property.

4. The number of features should be kept to a minimum to enable easier classifier training. Additionally, extracting fewer features can save time and effort. By satisfying the above three criteria, this criterion can be met.

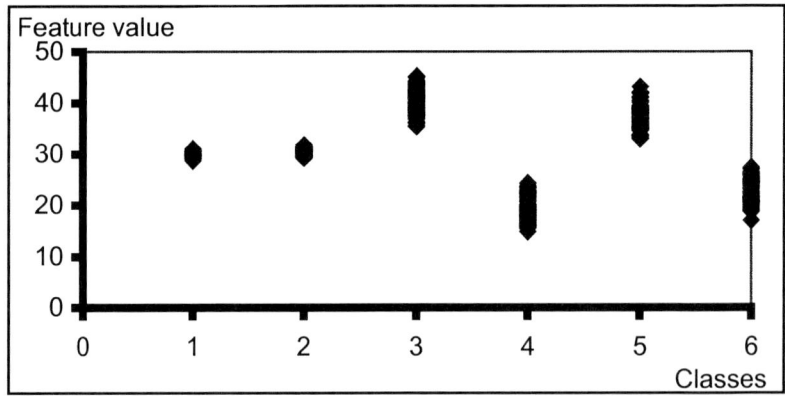

Figure 4.11 Values of one feature for different object classes

Figure 4.11 shows results on feature extraction from Alcock and Manolopoulos [1999]. Plotted in this figure are the values of one feature for 600 objects. The xaxis shows six different object classes while the y-axis displays the feature values. It can be seen that the feature has a very low intra-class variation for classes 1 and 2. Also, it has a relatively large inter-class variation between classes 3 and 4. The inter-class variation between classes 3 and 5 is low. Therefore, the feature would be good at discriminating certain classes from each other but would not be suitable to differentiate all classes on its own.

Figure 4.12 gives an example of two features that are approximately correlated. Here, the objects are of two types with one being represented by circles and the other by squares. It can be seen that as the value of one feature increases, the value of the other increases proportionately. Thus, only one of the two features is required for classification. The example given in the figure is a very simple correlation. Another simple correlation is where a feature is the reciprocal of a second feature. This is called inverse correlation. In more complicated examples of correlation, the value of a feature can be found as a function of another feature.

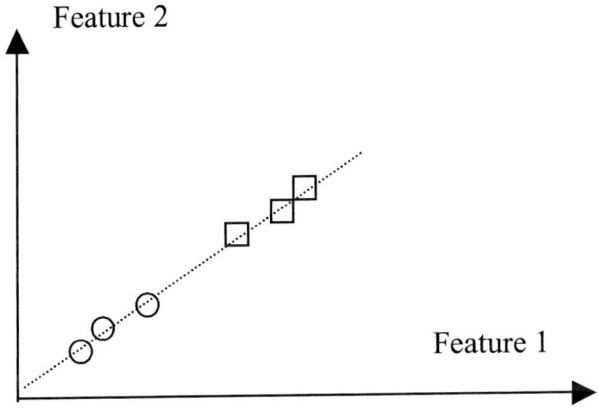

Figure 4.12 Approximately correlated features

Figure 4.13 illustrates the values of two features for two classes, again represented as circles and squares. Feature 2 is not a good feature to discriminate between the classes because given a value of feature 2, no reliable estimate of whether the object is a circle or a square can be made. On the other hand, feature 1 is a very good feature to differentiate between the two types because when its value is known, a reliable estimate of its type can be given.

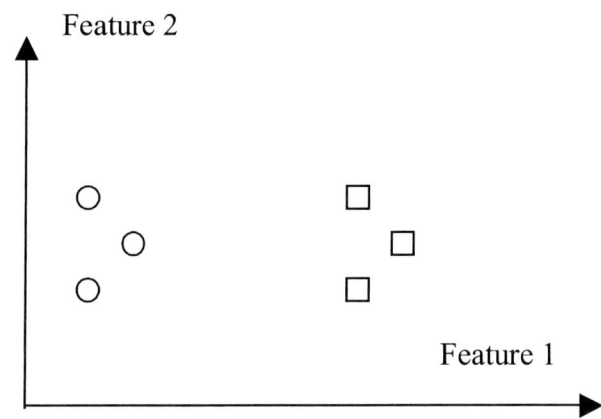

Figure 4.13 Good and bad features

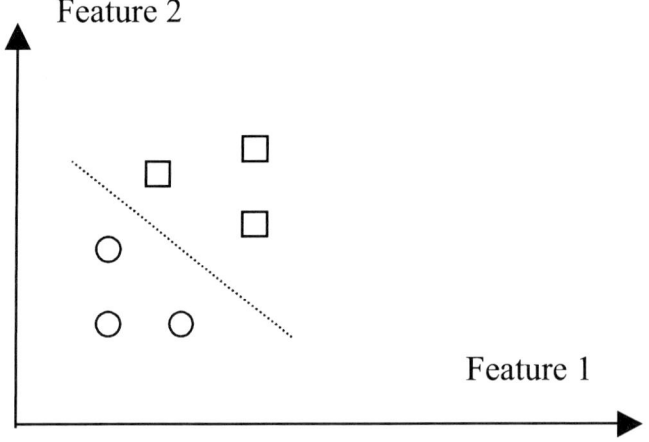

Figure 4.14 Two features required for classification

Figure 4.14 shows an example where two features are required for classification. When either feature is employed individually, the two classes cannot be distinguished. However, when both are utilised together, classification is straightforward.

Feature vectors must be evaluated to obtain a measure of their quality. Based on this, feature selection methods can be divided into two main approaches. The first is to evaluate the features based on statistical methods. The second is to determine the performance of the feature vectors on a given classifier.

4.4.1 Statistical Approach

Packianather and Drake [2000] carried out work on feature selection using statistical techniques. In their method, three measures were calculated, based on Castleman's criteria.

The equations for the three measures are given below. In the notation, there are n features to be analysed and feature j is labelled f_j where $1 \leq j \leq n$.

- intra-class variation:

$$iacv(f_j) = \frac{\sigma^2(f_j)}{\langle \sigma^2(f) \rangle}$$

(4.39)

where the numerator represents the variance of f_j and the denominator the mean variance of all features.

- inter-class variation:

$$ircv(f_j) = \sum_{i=1}^{i=n} \frac{|\mu(f_j) - \mu(f_i)|}{\sqrt{\sigma^2(f_j) + \sigma^2(f_i)}}$$

(4.40)

where $\mu(f_j)$ is the mean value for feature j.

- correlation:

$$cor(f_j, f_k) = \frac{\sum_{i=1}^{i=n} (v_{ij} - \mu(f_j))(v_{ik} - \mu(f_k))}{n\sigma(f_j)\sigma(f_k)}$$

(4.41)

where v_{ij} is the value of feature j for example i. The values for correlation are bounded by ±1. In practice, it will never take a value of 0. A high positive value indicates a large positive correlation, whilst a large negative value represents a high inverse correlation.

Features are rejected if their intra-class variation is above a given threshold or their inter-class variation is below a given threshold. Features are also rejected if they are highly correlated with another feature and are the worst performing of the two features.

The main advantage of the statistical approach is that the features selected are not classifier dependent. A problem is the determination of suitable thresholds.

Inductive learning has been incorporated into feature selection. Kamber et al. [1997] measured the performance of features by calculating their information gain. Information gain, or entropy, is employed in the ID3 and C4.5 inductive learning algorithms. Two methods can be adopted for accepting or rejecting features. First, the best X features can be kept for classification, where X is a user-specified parameter. Second, features can be rejected if they have an information gain below a given threshold. Another popular algorithm for building decision trees is CART [Breiman et al., 1984]. As decision trees put features at the top of the tree if they have a large information content, CART can also be employed for feature ranking and selection. The measure employed in CART is called node impurity. A node that contains patterns of only one type has a low impurity, whereas a node containing examples of many types will have a higher impurity.

There are many other measures to assess the quality of features. It is possible to use the probability density functions of each of the features [Devijver and Kittler, 1982]. A feature is considered optimal if its average class overlap is small. Pal and Chakraborty [1986] employed measures of set fuzziness for feature evaluation. Their measures were the indices of fuzziness, entropy and π–ness.

It should be noted that there is often some interaction between the features in the feature vector. A feature that performs poorly on its own may include valuable information that is only revealed when it is combined with another. Therefore, the statistical approach to feature selection may require a search method to find the optimal feature vector. Search techniques are usually required in the classification-based approach to feature selection and so will be described in the following section.

4.4.2 Classification-based Approach

The classification-based approach to feature selection is to test different subsets of the overall feature set on a particular classifier to determine which is the best one. In most cases, it is impossible to test all possible combinations. When there are n features, the possible number of subsets of those features is 2^n-1. Thus, even for relatively small values of n, the number of subsets becomes prohibitively large. To confront the large number of subsets, it is necessary to have an efficient search technique. The most common methods for this are conventional techniques and genetic algorithms. There have also been several attempts to utilise neural networks to search for optimum feature sets.

Siedlecki and Sklansky [1988] and Webb [1999] have given reviews of conventional search techniques for feature selection. These include strategies such as forward, backward, bi-directional and branch-and-bound searching. Forward search starts with an empty feature set. Then, the best feature is added to this set and the performance tested. Adding features terminates either when a pre-specified number of features is reached or adding the new feature does not improve performance. Backward searching operates in the opposite manner. It starts with the total set of features and the worst feature is removed. In an iterative manner, features are removed until doing so makes performance worse. The advantage of forward search over backward search is that training takes less time as it is normally working with fewer features than backward searching. Bi-directional search combines forward and backward searching. The branch-and-bound technique relies on the monotonicity criterion. The criterion states that if there are two feature sets, X_1 and X_2, and $P(X)$ is the performance of set X then:

$$X_1 \subset X_2 \Rightarrow P(X_1) < P(X_2)$$

In the branch and bound technique, if a subset X_1 has a performance of less than a specified threshold then no subsets of X_1 are evaluated. In this way, the number of subsets to be tested reduces. There are two main problems with the branch-and-bound algorithm. First, the monotonicity criterion is very rarely satisfied. Second, if the chosen threshold is too low then the search space is not reduced significantly.

Oyeleye and Lehtihet [1998] selected features for solder joint inspection using backward search. The selected features were found to be effective for classifying solder defects. Pham and Alcock [1999] also adopted a backward search method. In their technique, a MLP neural network was chosen as the performance evaluator. First, trials were performed leaving out one feature each time. Each subset was executed three times and the average performance taken. In the experiments, the thirty-one features were labelled F1 to F31. It was found that leaving out feature F10 or F14 improved the performance (Table 4.2). Next, leaving out combinations of the worst seven features was evaluated. It was expected that leaving out F10 and F14 together would improve the performance further. However, none of the combinations were superior to leaving out just F10. Two explanations were suggested. First, averaging the results over just three runs may not be enough. Second, the technique does not take into consideration interactions between the features.

Features left out	Run 1	Run 2	Run 3	Average
F10	87.5%	89.8%	90.9%	89.4%
F14	87.5%	87.5%	89.8%	88.3%
F10, F1	85.2%	90.9%	85.2%	87.1%
F14, F27	86.4%	85.2%	88.6%	86.7%
F10, F27	80.7%	92.0%	86.4%	86.4%
F14, F21	81.8%	93.2%	83.0%	86.0%
None	*86.4%*	*81.8%*	*87.5%*	*85.2%*
F10, F14	88.6%	81.8%	83.0%	84.5%
F10, F20	78.4%	85.2%	87.5%	83.7%
F14, F1	87.5%	80.7%	83.0%	83.7%
F14, F20	81.8%	90.9%	78.4%	83.7%
F10, F21	84.1%	83.0%	84.1%	83.1%
F14, F11	75.0%	87.5%	86.4%	83.0%
F10, F11	79.5%	84.1%	83.0%	82.2%
F10, F14, F20, F1, F11, F21, F27	78.4%	78.4%	87.5%	81.4%
F10, F14, F20	79.5%	75.0%	83.0%	79.2%

Table 4.2 Effect on classification performance of leaving out feature combinations

The performance of a classifier with a given subset of features can take a range of values. For example, if a neural network is trained using a given subset of features, the performance obtained will vary from one training session to the next. Kupinski and Giger [1999] determined that the variation in performance of a given subset of features depends on the training data set size, the number of initial features, the number of features selected and the performance of the true optimal feature set. With small training sets and a large number of features from which to choose, a large variation in performance was observed. To overcome this variation, several tests should be carried out and the average result calculated. However, there is a trade-off as more experiments will give more accurate results but have a longer execution time. Belue and Bauer [1995] suggested that the classifier would need to be trained as many as thirty times on each feature subset to obtain a reliable confidence interval in its performance. If the network needs to be trained so many times for each feature subset, then it is critical that the classifier used for evaluation operates as quickly as possible.

One advantage of employing genetic algorithms for feature selection is that they do not require the features to obey the monotonicity criterion. Several researchers have tested GAs to search for optimum feature sets [Punch et al., 1993; Sahiner et al., 1996; Siedlecki and Sklansky, 1989]. Estevez et al. [1999] also implemented this approach. Feature subsets were represented as binary strings, where a '1' in position i indicates that feature i is present and a '0' in that position shows that it is absent. A niching technique called *deterministic crowding* was employed. In this method, chromosomes are randomly combined and the resulting chromosomes compete with their neighbour using Hamming distance. The best chromosomes are kept for the next generation. To evaluate the fitness of a chromosome, the selected feature subset was fed into a neural network. The aim of the fitness function was to maximise the accuracy of classification whilst minimising the number of features. The equation for the fitness function was:

$$\text{fitness} = a - \lambda * (n / N) \tag{4.42}$$

where a is the accuracy of classification achieved with the feature subset, λ is a user-specified parameter, n is the number of features in the subset and N is the total number of features. The value for λ in the experiments was set to 0.01.

The calculation of a is the computationally-intensive part of the process. Additionally, with neural networks, different values for accuracy are obtained each time the network is trained. Therefore, for each configuration, the networks were trained three times and the best performance was recorded. To save time, performances for chromosomes were recorded so that if the same chromosome recurs, the accuracy does not need to be recalculated. Additionally, a new mutation operator was employed to speed up the convergence of the algorithm. The operator was based on the SSM (Statistical Stepwise Method) weight elimination technique that removes small weights during neural network training. This method was extended to GAs to enable the mutation operation to remove poor features.

To test the GA selection method, an image database was built up. The images were segmented and nine hundred defects were found. This included one hundred examples of each of nine defect types. Seventy-two features were extracted from each object. These included twenty-four geometrical features and twelve features extracted from each of four colour components (red, green, blue and grey scale). The GA was able to reduce the number of features from seventy-two to twenty-one whilst simultaneously increasing the performance from 88.6% to 91.2%.

Bril et al. [1992] also employed GAs for feature selection. Their work involved three innovative aspects. First, Genetic Algorithms with Punctuated Equilibria

(GAPE) were used. These are different from standard GAs in that they operate with several populations separately and combine the results at the end. Second, instead of utilising a MLP neural network to calculate the evaluation functions, they incorporated a counter-propagation (CP) network. The CP network is closely related to the nearest-neighbour classifier and so has a relatively short execution time. This meant that the GA could search in a reduced time. Third, training set sampling was carried out. This reduced the number of distance calculations required by the CP network.

Several researchers have employed neural networks to determine the optimum feature set [Priddy et al., 1993; Steppe et al., 1996]. The saliency measure of a feature is the sum of the squared weights of the input neuron for that feature. The inputs that are more useful in classification will have a larger saliency value. Belue and Bauer [1995] added an extra input to the neural network that just consisted of noise. Then, the significance of each feature was determined by measuring its saliency. Any feature that had a saliency approximately equal to or less than the noise input was then considered useless. Setiono and Liu [1997] tested each feature in turn by making the weights from the feature's input neuron equal to zero. Then, the difference between the network's accuracy before and after the change was calculated. The input feature with the lowest accuracy was then removed. The process of testing all the features was repeated with this feature removed. Meyer-Base and Watzel [1998] utilised the RBF neural network for feature selection. By changing slightly the three-layer architecture of a RBF network, it was possible to obtain a measure to rank all the features.

Steppe et al. [1996] found that the features with the highest saliency values were not always the best subset of features. This is because one feature may interact with another feature to give additional information. When one of the features is removed, it is not only the value of the feature that is lost but also the value of its interactions with other features.

4.4.3 Other Approaches

To overcome the problem of feature vector design and selection, Lampinen and Smolander [1996] employed a SOFM to determine automatically features from windows in the image. The pixel values from a window in the image are first passed through a Gabor filter before being presented to the SOFM. Then, the SOFM clusters similar pixel distributions together. Finally, the output of the SOFM is presented to a MLP for classification. One problem with this technique is that very large training sets can be produced from each image and it is important to select a subset of these that accurately represents the whole set. Also, it remains to be proved whether features obtained using this method produce better results than those chosen conventionally.

Ramze-Rezaee et al. [1999] employed fuzzy logic in their feature selection method. First, the feature values are converted to fuzzy variables. Next, conventional search techniques determine the best feature set. In the experiments, three benchmark data sets and one image analysis problem were employed. The benchmark data sets were the IRIS, DIABETES and VEHICLES data sets, which are now publicly available on the Internet [Blake et al., 2002]. It was found that with a reduced number of fuzzy features, a better classification performance could be obtained than with the original data sets.

The feature selection problem assumes that the best set of features is a subset of the known features. However, generating extra features using mathematical combinations of existing features can give better performance than any subset of the original features. Perantonis and Virvilis [1999] worked on generating new features using a technique similar to Principal Components Analysis [Webb, 1999]. The newly generated features were presented to a MLP for classification and it was found that they gave a higher performance. The problem of how best to combine the features is similar to the feature selection problem in that there are a very large number of combinations that can be made and the optimal set must be searched for.

4.5 Discussion

It has been shown in this chapter that the selection of an optimum set of features has a considerable effect on classification performance. However, determining suitable features for a given application is often problematic. It is suggested that in the early stages of feature set development, a large number of features be derived. These can be found by reading literature on similar applications and also by "brainstorming". When a number of possible features has been identified, work can begin to determine which of these features are beneficial for classification and which are detrimental. As a preliminary stage in feature selection, manual statistical analysis should be performed to identify possibly redundant features. Then, an automatic search technique can be used to reduce the feature set further. Currently, an active research area is to find the most suitable technique for automatic feature selection.

4.6 Summary

After segmentation has been performed, a segmented image is generated, containing objects. For image understanding, it is necessary to know what these objects represent. Typically, features are extracted from each object, which are then presented to a classifier. Important issues in feature extraction are to determine the features to employ and to eliminate unnecessary features. Derivation of the optimal feature vector plays a key role in obtaining a high classifier performance.

For automated selection of features, there are two main approaches. First, the features can be assessed using statistical measures. Second, their performance can be determined based on their classification performance. In either case, it is best to search through a large number of feature subset combinations, as a feature that gives a bad performance on its own may give good results when combined with other features. One search method that has given promising results is genetic algorithms.

References

Alcock R.J. and Manolopoulos Y. (1999) Genetic Algorithms for Inductive Learning. In *Computation Intelligence and Applications*. (ed. Mastorakis N.) World Scientific, Singapore. pp. 85 - 90.

Belue L.M. and Bauer K.W. (1995) Determining Input Features for Multilayer Perceptrons. *Neurocomputing*. Vol. 7, pp. 111 - 121.

Blake C., Keogh E. and Merz C.J. (2002) *UCI Repository of Machine Learning Databases*. University of California, Department of Information and Computer Science, Irvine, CA. http://www.ics.uci.edu/~mlearn/MLRepository.html

Breiman L., Friedman J.H., Olshen R.A. and Stone C.J. (1984) *Classification and Regression Trees*. Wadsworth and Brooks. Belmont, CA.

Bril F.Z., Brown D.E. and Worthy N.M. (1992) Fast Genetic Selection of Features for Neural Network Classifiers. *IEEE Trans. on Neural Networks*. Vol. 3, No. 2, pp. 324 - 328.

Brunner C.C., Maristany A.G., Butler D.A., Van Leeuwen D. and Funck J.W. (1992). An Evaluation of Colour Spaces for Detecting Defects in Douglas-fir Veneer. *Industrial Metrology*. Part 2, pp. 169 - 184.

Castleman K.R. (1979). *Digital Image Processing.* Prentice Hall, Englewood Cliffs, NJ.

Conners R.W. and Harlow C.A. (1980) A Theoretical Comparison of Texture Algorithms. *IEEE Trans. on Pattern Analysis and Machine Intelligence.* Vol. 2, No. 3, pp. 204 - 222.

Conners R.W., McMillin C.W., Lin K. and Vasquez-Espinosa R.E. (1983) Identifying and Locating Surface Defects in Wood: Part of an Automated Lumber Processing System. *IEEE Trans. on Pattern Analysis and Machine Intelligence.* Vol. 5, No. 6, pp. 573 - 583.

Devijver P.A. and Kittler J. (1982) *Pattern Recognition: A Statistical Approach.* Prentice Hall. Englewood Cliffs, NJ.

Estevez P.A., Fernandez M., Alcock R.J. and Packianather M.S. (1999) Selection of Features for the Classification of Wood Boards. *Proc. 9th Int. Conf. on Artificial Neural Networks.* Edinburgh, UK. pp. 347 - 352.

Gonzalez R.C. and Woods R.E. (1992) *Digital Image Processing* (3rd ed.) Addison-Wesley, Reading, MA.

Haralick R.M., Shanmugam K. and Dinstein I. (1973). Textural Features for Image Classification. *IEEE Trans. on Systems, Man and Cybernetics.* Vol. 3, No. 6, pp. 610 - 621.

Kamber M., Winstone L., Gong W., Cheng S., and Han J. (1997) Generalization and Decision Tree Induction: Efficient Classification in Data Mining. *Proc. Int. Workshop on Research Issues on Data Engineering.* Birmingham, UK, pp. 111-120.

Kupinski M.A. and Giger M.L. (1999) Feature Selection with Limited Datasets. *Medical Physics.* Vol. 26, No. 10, pp. 2176 - 2182.

Lampinen J. and Smolander S. (1996) Self-Organising Feature Extraction in Recognition of Wood Surface Defects and Color Images. *Int. Journal of Pattern Recognition and Artificial Intelligence.* Vol. 10, No. 2, pp. 97 - 113.

Meyer-Base A. and Watzel R. (1998) Transformation Radial Basis Neural Network for Relevant Feature Selection. *Pattern Recognition Letters.* Vol. 19, No. 14, pp. 1301 - 1306.

Oyeleye O. and Lehtihet E.A. (1998) A Classification Algorithm and Optimal Feature Selection Methodology for Automated Solder Joint Defect Inspection. *Journal of Manufacturing Systems*. Vol. 17, No. 4, pp. 251 - 262.

Packianather M.S. and Drake P.R. (2000) Neural Networks for Classifying Images of Wood Veneer. Part 2. *Int. Journal of Advanced Manufacturing Technology*. Vol. 16, No. 6, pp. 424 - 433.

Pal S.K. and Chakraborty B. (1986) Fuzzy Set Theoretic Measure for Automatic Feature Evaluation. *IEEE Trans. on Systems, Man and Cybernetics*. Vol. 16, No. 5, pp. 754 - 760.

Perantonis S.J. and Virvilis V. (1999) Input Feature Extraction for Multilayered Perceptrons using Supervised Principal Component Analysis. *Neural Processing Letters*. Vol. 10, No. 3, pp. 243 - 252.

Pham D.T. and Alcock R.J. (1999) Automated Visual Inspection of Wood Boards: Selection of Features for Defect Classification by a Neural Network. *Proc. IMechE. Part E - Journal of Process Mechanical Engineering*. Vol. 213, No. E4, pp. 231 - 245.

Pham D.T. and Cetiner B.G. (1996) A New Method for Describing Texture. *Proc. 3rd Int. Workshop on Image and Signal Processing on Advances in Computational Intelligence*. Manchester. (eds. Mertzios B.G. and Liatsis P.) Elsevier Science. pp. 187 - 190.

Priddy K.L., Rogers S.K., Ruck D.W., Tarr G.L. and Kabrisky M. (1993) Bayesian Selection of Important Features for Feedforward Neural Networks. *Neurocomputing*. Vol. 5, pp. 91 - 103.

Punch W.F., Goodman E.D., Pei M., Chia Shun L., Hovland P. and Enbody R. (1993) Further Research on Feature Selection and Classification using Genetic Algorithms. *Proc. 5th Int. Conf. on Genetic Algorithms*. University of Illinois, Urbana-Champaign, IL. Vol. 5, pp. 557 - 564.

Ramze-Rezaee M., Goedhart B., Lelieveldt B.P.F. and Reiber J.H.C. (1999) Fuzzy Feature Selection. *Pattern Recognition*. Vol. 32, pp. 2011 - 2019.

Russ J.C. (1995) *The Image Processing Handbook*. (2nd ed.). CRC Press, Boca Raton, FL.

Sahiner B., Chan H., Wei D., Petrick N., Helvie M.A., Adler D.D. and Goodsitt M.M. (1996) Image Feature Selection by a Genetic Algorithm: Application to

Classification of Mass and Normal Breast Tissue. *Medical Physics*. Vol. 23, No. 10, pp. 1671 - 1684.

Setiono R. and Liu H. (1997) Neural Network Feature Selector. *IEEE Trans. Neural Networks*. Vol. 8, No. 3, pp. 654 - 662.

Siedlecki W. and Sklansky J. (1988) On Automatic Feature Selection. *Int. Journal of Pattern Recognition and Artifical Intelligence*. Vol. 2, No. 2, pp. 197 - 220.

Siedlecki W. and Sklansky J. (1989) A Note on Genetic Algorithms for Large-Scale Feature Selection. *Pattern Recognition Letters*. Vol. 10, pp. 335 - 347.

Sobey P.J. and Semple E.C. (1989) Detection and Sizing Visual Features in Wood Using Tonal Measures and a Classification Algorithm. *Pattern Recognition*. Vol. 22, No. 4, pp. 367 - 380.

Sobey P.J. (1990) Automated Optical Grading of Timber. *SPIE Vol. 1379: Optics in Agriculture*. pp. 168 - 179.

Song K.Y., Petrou M. and Kittler J. (1992) Texture Defect Detection: A Review. *SPIE Vol. 1708: Applications of Artificial Intelligence X: Machine Vision and Robotics*. Orlando, FL. pp. 99 - 106.

Steppe J.M., Bauer K.W. and Rogers S.K. (1996) Integrated Feature and Architecture Selection. *IEEE Trans. Neural Networks*. Vol. 7, No. 4, pp. 1007 - 1014.

Webb A. (1999) *Statistical Pattern Recognition*. Arnold, London.

Weszka J.S, Dyer C.R. and Rosenfeld A. (1976) A Comparative Study of Texture Measures for Terrain Classification. *IEEE Trans. on Systems, Man and Cybernetics.* Vol. 6, No. 4, pp. 269 - 285.

Problems

1.   Calculate the mean and skewness for the following histograms:

x	$H_1(x)$	$H_2(x)$
0	10	40
1	20	30
2	20	20
3	10	10

Comment on the results.

2. Calculate the co-occurrence matrix for the following quantised image segment, using d=1 and θ=0°:

0	1	1	1
0	1	2	2
0	1	2	3
0	1	2	3

3. Determine which of the following features are approximately correlated:

	Feature 1	Feature 2	Feature 3
Object 1	0.9	2.9	-2.0
Object 2	2.1	2.1	-3.9
Object 3	3.0	0.9	-6.1

4. Given that the three objects in Problem 3 belong to the same class, calculate their intra-class variation.

5. Discuss the intra and inter-class variations of the following feature:

Feature value	Class
4	1
8	1
6	1
5	2
7	2
6	2
1	3
2	3
1	3

Would this feature be enough on its own for classification?

6. Discuss the suitability of the following two features for classifying between circles, squares and triangles.

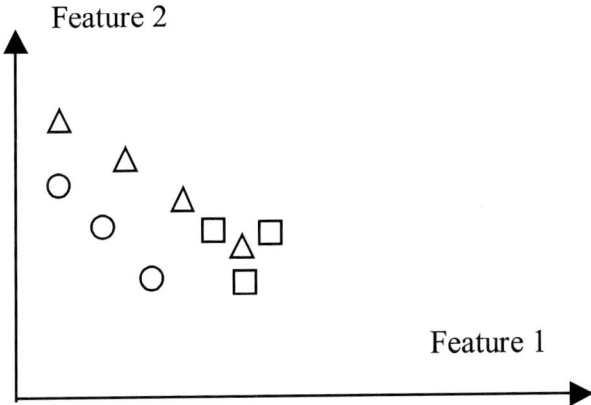

7. In ProVision, the *W Blob Analysis* function can be employed to extract features from binary images. Open the DATE project and list the features used in this program.

8. By using the *Test* and *Display* buttons in the *W Blob Analysis* dialog box, how many blobs are found for the first image of the DATE project and in what order are they sorted?

Chapter 5

Classification

In classification, a feature vector is taken and the type, or class, of the object that the vector represents is determined. Figure 5.1 shows an input-output model of a classifier. The input to the classifier is n features $(x_1, x_2, ..., x_n)$. The classifier processes these inputs in some way and gives outputs. The outputs can be numbers $t_1, t_2, ..., t_m$, where there are m output types. The overall output of the classifier is type i if $t_i > t_j$ for $1 \geq i,j \geq m$, $i \neq j$.

Figure 5.1 Input-output model of a classifier

A simple example of a classifier is a system to differentiate between oranges and mandarins. Just one feature would be required, namely, the size of the object. Then, if output t_1 represented oranges and t_2 mandarins, output t_1 of the classifier should give a larger output than t_2 when an orange is presented to the system.

For analysis of classification, confusion matrices are often used. A confusion matrix indicates into which classes the examples of each type have been classified. Figure 5.2 gives an example of a confusion matrix. Confusion matrices show the classes are confused with one other. This information can be utilised to improve feature extraction or classification.

Many different classifiers have been used to categorise objects in AVI images. These can be divided into three main categories: statistical classifiers, rule-based systems and neural networks.

		Classified type		Total
		Mandarins	Oranges	**Total**
Actual	Mandarins	40	10	50
type	Oranges	1	49	50
	Total	41	59	100

Figure 5.2 Confusion matrix for mandarin and orange classification

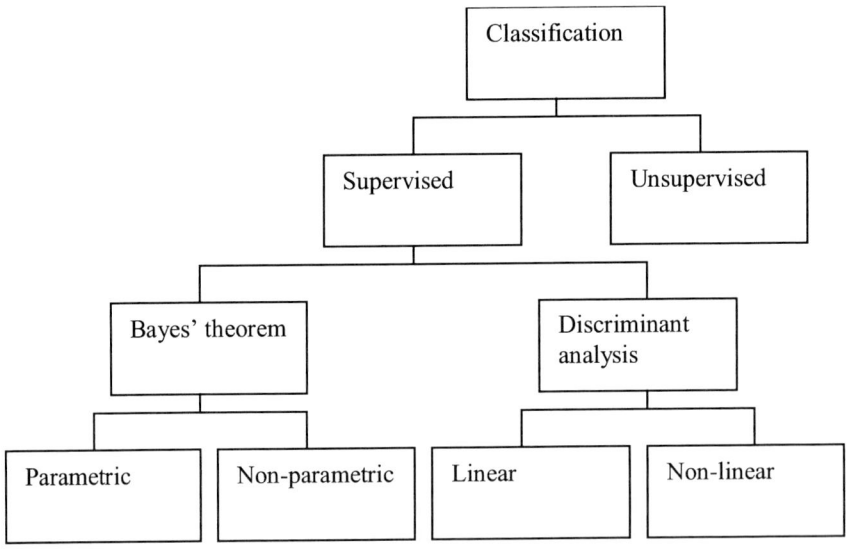

Figure 5.3 Categorisation of classification techniques

Figure 5.3 gives a categorisation of classification techniques [Webb 1999]. As mentioned in Chapter 1, classifiers are either supervised or unsupervised. Supervised classifiers have a training set with inputs and associated classes whilst unsupervised classifiers utilise only input data. Unsupervised classification operates by grouping together inputs that are similar and so is also called clustering.

Unsupervised classifiers include the neural networks ART and SOFM, as well as established algorithms such as k-means clustering. Supervised classification methods are divided into classifiers based on Bayes' theorem and those using discriminant analysis.

Classifiers based on Bayes' theorem are further subdivided into parametric and non-parametric classifiers. Parametric classifiers require the specification of the parameters of the training data distribution. The Bayesian classifier is an example of a parametric classifier. Non-parametric classifiers do not require parameters to be specified. An example of this type of classifier is the minimum-distance classifier.

Discriminant classifiers are divided into linear and non-linear types. In two-dimensional terms, classification can be performed using a linear classifier if a straight line can be drawn between the classes. Examples of linear classifiers are rule-based systems and tree classifiers. A non-linear classifier is required if a non-linear boundary between the classes is required. Currently, a common non-linear classifier is the MLP neural network.

Classification requires a training set for the development of the classifier. The set of training examples should be representative of the population of all examples. A good training set should have two properties. First, it should be large enough to cover the complete range of examples that would be encountered during operation. Second, the examples in the training set should be of a good quality. In other words, noisy examples should be removed. According to Davies [1996], the number of examples of each type in the training set should not be the same. Rather, it should depend on the natural frequency of occurrence of the corresponding types. Also, it is a good practice to randomise the order of the training examples. The result of some classifiers depends upon the order of the training set. Thus, it is better to have examples of the same type spread throughout the training set.

To evaluate the developed classifier, a test data set is required. The test set contains examples that have not been shown to the classifier during training. The purpose of the test set is to evaluate how well the classifier will perform during normal operation. A classifier that has a good test set accuracy has good generalisation ability.

5.1 Bayes' Theorem Classifiers

Bayes' theorem classifiers include Bayesian, minimum-distance and K-Nearest Neighbour (KNN) classifiers.

The Bayesian classifier is a popular classifier [Webb, 1999]. It is based on Bayes' theorem of conditional probabilities. In terms of classification using one feature, this theorem states:

$$P(t_i \mid v) = \frac{P(v \mid t_i)P(t_i)}{P(v)} \tag{5.1}$$

where $P(t_i|v)$ is the probability of the example being of type t_i if the feature has value v, $P(v|t_i)$ is the probability of the feature being value v given that the object type is t_i, $P(t_i)$ is the probability of type t_i occurring and $P(v)$ is the probability that the feature will have value v. The values of the terms on the right-hand side of the equation are obtained from the training examples. With the Bayesian classifier, an object whose feature has value v will be deemed to belong to type t_i if $P(t_i|v)$ is larger than the probability associated with any other class.

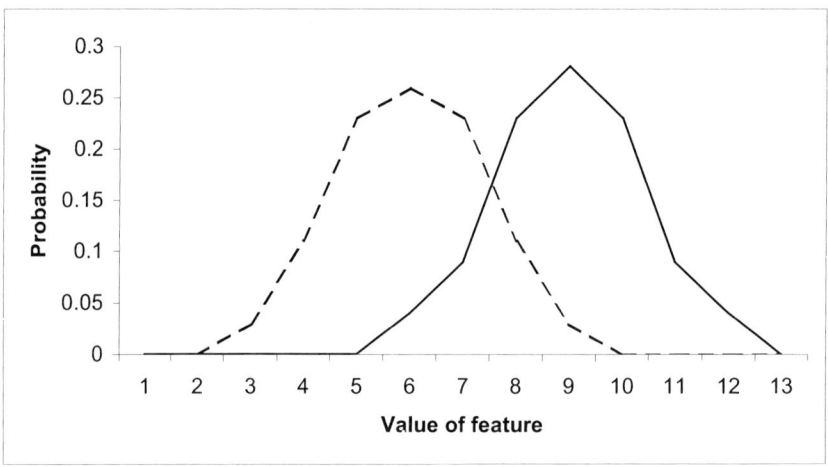

Figure 5.4 Probability density functions

Figure 5.4 illustrates a simple example of employing the Bayesian classifier in a problem involving one feature and two classes. Probability density functions (PDFs) are shown for the two classes. Given the value of the feature, the probability of each class can be obtained. Then, the actual class is the one that yields the highest probability for that feature value. The error of the classifier on the training set is the overlap between the two PDFs. To reduce the classification error, more features can be added.

The Bayesian classifier is an optimal classifier in the sense that it minimises the probability of classification errors. However, for this to be true, it is necessary for each PDF to have the correct distribution. It is common practice to assume that the PDFs are Gaussian. Clearly, the Bayesian classifier is not the optimum classifier when the class distributions are not truly Gaussian.

An example of an industrial AVI application of Bayesian classifiers is provided by the bottle inspection work of Magee et al. [1993] who chose a Bayesian classifier with Gaussian functions. Images were derived of the tops of the bottles and five features extracted for classification.

A minimum-distance (MD), or nearest-neighbour, classifier is a simple classifier that determines the type of a new object by measuring the distance of its features from those of objects for which a type is already known. The distance D between two feature vectors **r** and **s**, both with n features, can be calculated in several different ways. One of the most common is the Minkowski distance:

$$D = \sum_{i=1}^{n} ((r_i - s_i)^p)^{1/q}$$

(5.2)

When p=q=2, the formula represents the popular Euclidean distance. Larger values of p and q emphasise greater distances but are rarely used.

In the MD classifier, the unknown object is said to be of the same type as the known object to which it is found to be closest to. The MD classifier can be improved by adding weightings to each of the features, as some features may be more important than others. A problem with the MD classifier is that if all training patterns are stored then a large amount of memory will be required and the execution time will be long. This can be improved by pruning examples that have little effect on the overall classification.

The k-nearest-neighbour (KNN) classifier is an extension of the minimum-distance classifier. However, instead of just finding the closest feature vector to that of the unknown feature, K nearest neighbours are found. Typically, K is an odd number such as 3 or 5. Then, the type that occurs most frequently among the K feature vectors is the type assigned to the unclassified feature vector.

Figure 5.5 illustrates the operation of the MD and KNN classifiers. If it were required to classify an object with feature values represented by the cross then using the MD classifier, it would be considered to be of the same type as the circle, as this is closest to the cross. However, using a KNN classifier with K equal to three, the three closest points would be two squares and one circle. Thus, the cross would be considered to be of the same type as the squares.

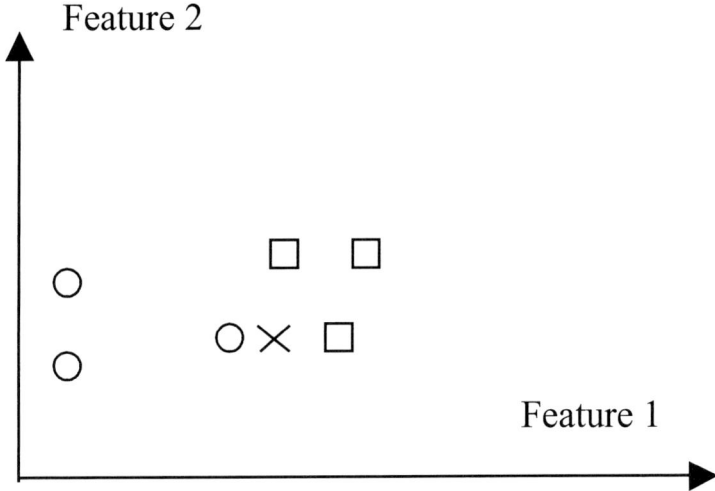

Figure 5.5 Distance-based classification

As with all classifiers, it is important that the training set is representative of all examples that might be faced in the test set. Therefore, a large enough training set is required. However, if the training set becomes too large then some method of example selection is needed.

5.2 Rule-Based Classification

Rule-based systems classify according to if-then rules. The following is an example of a rule that might be employed in the inspection of round objects:

IF circularity < 0.9 THEN reject

This rule would reject objects with a circularity value less than 0.9.

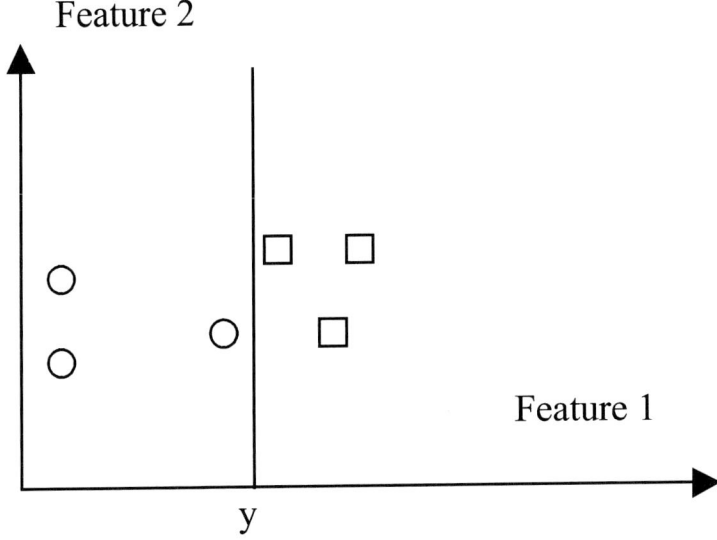

Figure 5.6 Illustration of the operation of a simple rule-based classifier

Figure 5.6 shows the operation of a simple rule for the classification of the objects in Figure 5.5. A rule can be developed to classify between the circles and squares:

IF feature 1 > y THEN type is square

Whether this rule would give a good performance or not depends on how good a representation the six given examples are of the total set of examples that could occur for the two types. Because a straight line can be drawn between the different classes, the problem is said to be linearly separable and only a simple linear classifier is needed. In three-dimensional space, the problem is linear if the classes can be separated using planes. For more complex problems, straight lines cannot be drawn between the classes and non-linear classifiers would be required.

To test whether rules could be employed for defect identification, the authors developed a rule-based classification system [Pham and Alcock, 1999a]. Only one rule per defect type was allowed to try to ensure good generalisation. Based on a training set comprising twenty-seven features and ninety-five examples, rules could be developed manually for only ten of the features because no relationship was apparent for the other seventeen features and the defect types. The process of deriving rules was time consuming and the resulting rules could only correctly classify less than 74% of the test set. It was found that there are several problems with manually deriving rules for classification:

1. Rules can only be derived for features that are meaningful to humans. For example, if a feature is obtained to measure the number of very dark pixels in an object then a rule for dark object types can be found from this. However, it is difficult manually to derive rules from features, such as kurtosis, which are hard to visualise.

2. Rules create crisp boundaries between classes in the feature space. For example, using the circularity rule the object would be rejected if its circularity were 0.899. Clearly, the feature value is very close to that for an accepted item.

3. Because rules are written using a training set, there is a tendency for the writer of the rules to optimise them on that set. To improve training set performance, it is possible to write many rules, each one only classifying a small part of the training set. The performance obtained with those rules would not be as good on cases not covered by the training set.

4. For certain object types, it is difficult to write rules. Further problems arise if the segmentation is imperfect and the detected objects do not match the actual objects. Therefore, rules about the shape and size of objects will not hold true on the segmented image.

5. Certain object types have very similar features and therefore are difficult to separate using rules.

Currently, there are a number of established algorithms to derive crisp rules. As mentioned in Chapter 1, these algorithms, called inductive learning algorithms, extract a rule set automatically from the training data. To test the applicability of inductive learning, the authors adopted the RULES-3 Plus inductive learning algorithm [Pham and Dimov, 1997]. This algorithm has proved effective on a number of classification tasks. However, a large number of rules (49 rules) were produced for the training set. Many rules were derived which were specific to just one training example. One rule covered 40 examples, of which only 23 were correctly classified. Thus, the generalisation ability of the rule set was limited.

The major advantage of a rule-based classifier is its transparency. When a classification is made, it can be seen which rules were involved to reach the decision. Also, because the rules can be seen, it is easy for them to be changed. This can also be a disadvantage because the user may include poor rules or too many rules. One of the major problems is how to obtain effective rules. Even with the aid of modern inductive learning programs, the task is not simple.

5.2.1 Fuzzy Rule-Based Classification

To overcome the problem of creating crisp boundaries between classes, fuzzy rule-based induction methods can be used. A fuzzy rule-based system requires a set of membership functions that relate feature values to classes. Figure 5.8 shows the size of an object and the membership functions for three different class types. In the example, if the size of the object to be classified is 16mm then it can be seen that it belongs to class one with a membership of 0.3 and to class 2 with a membership of 0.6. Thus, the output would be class 2.

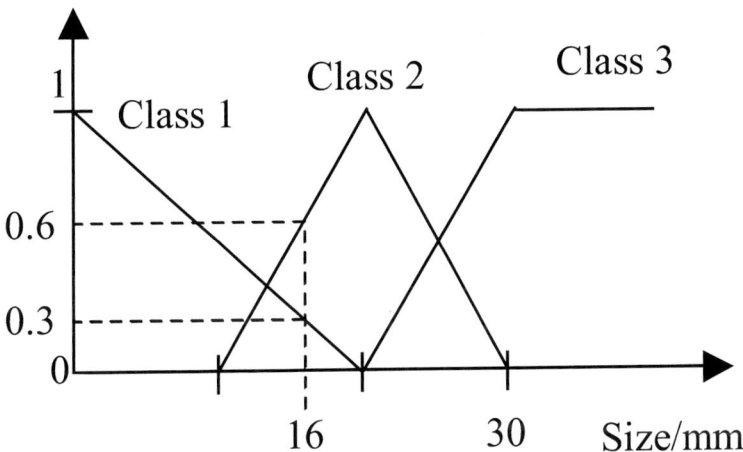

Figure 5.7 Fuzzy rule-based classification

In a normal fuzzy classifier, several features would be utilised. This means that the degree of membership for each class and feature must be combined to give an overall result. The simplest method of performing this would be to add the degrees of membership. Then, the class with the largest summed value would be the overall output type.

One major problem with fuzzy classification is how to determine the fuzzy membership functions. As mentioned in Chapter 1, there are commercial tools available that can automatically induce a fuzzy rule set from an input-output data set. Recent research developments on fuzzy rule induction can be found in [Ravi and Zimmermann, 2000; Shen and Chouchoulas, 2000].

5.3 MLP Neural Network Classification

As previously mentioned, one of the most common neural networks is the Multi-Layer Perceptron (MLP). The network comprises a number of networked artificial neurons. The MLP generates a mapping from a set of inputs to a set of outputs. In the case of AVI classification, the inputs are a feature vector and the output is the defect type.

Normally, features are scaled between ±1 or 0 and 1 to be used as MLP network inputs. The most common activation functions are linear for the input neurons and sigmoidal for the hidden and output neurons. Sigmoidal activation functions give outputs limited by 0 and 1. For classification in AVI, the network normally has as many input neurons as features and as many output neurons as there are object types. The user must choose the number of hidden neurons. During training, if an object is of type i then output i of the network is forced to a high value and all the other outputs to a low value. A high output is taken to be above 0.9 and a low output below 0.1 because with the sigmoidal activation function, outputs of 1 and 0 cannot be obtained without inputs of infinite magnitude to the neurons. When the network is operated in recall mode, the output with the highest activation is taken to be the output of the network. Initially, all weights in the network are randomly assigned values between ±W, where W is typically a number between 0 and 1.

A MLP trained by the back-propagation technique has five parameters to be set by the user. These are:
- the number of hidden layers;
- the number of neurons in each hidden layer;
- the learning rate η;
- the momentum α;
- the stopping criterion.

There is currently no widely-accepted technique for automatically setting the parameters. Experiments usually must be performed on the neural network to determine the effect of different settings on its performance. However, it is impossible to test every combination of parameter settings to find the optimum because α and η are continuous variables and the numbers of hidden layers and neurons could take any positive integer value.

However, according to the neural network literature [DTI, 1994], more than one hidden layer is rarely needed. The number of hidden neurons NH in the hidden layer is frequently stated to be dependent upon the number of inputs NI and the number of outputs NO of the network. The following equation has been suggested for determining the number of hidden neurons [DTI, 1994]:

$$NH = \frac{NI + NO}{2}$$

(5.3)

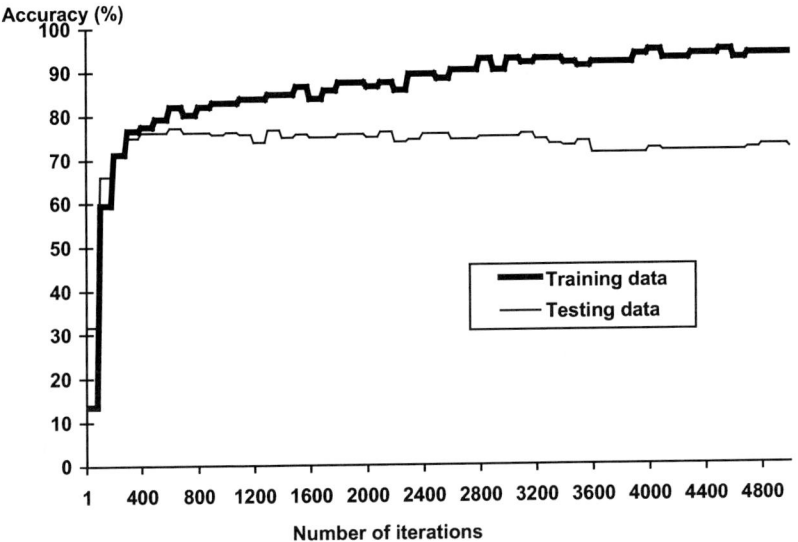

Figure 5.8 Training and test set performances of the MLP

When to stop the training, to ensure good generalisation, is another important issue. The rule that determines when training should be stopped is called the stopping criterion. In the early stages of training, the network learns the general properties of the training data. Subsequently, the network learns the specific properties of the training data and thus it also starts to learn the noise in the training set. Figure 5.8 shows the performances for a training set and a test set plotted against the number of training iterations. One iteration is completed when all examples in the training set have been presented to the network and the weights of all the connections in the network have been changed once. It can be seen that after a certain point, the performance on the test set deteriorates but that on the training set continues to improve. To achieve good generalisation, an efficient stopping criterion should be selected, which can be achieved by employing a validation set. For this, the examples are divided into three sets: training, test and validation. The network is trained using the training data and, after a fixed number of iterations, is tested using the validation data. If there is no improvement in the performance on the validation set after a pre-determined number of iterations then the training can be terminated.

Pham and Alcock [1999b] performed experiments to determine optimal values for NH, η and α for the task of classifying wood defects. The results are shown in Figs 5.9 to 5.11.

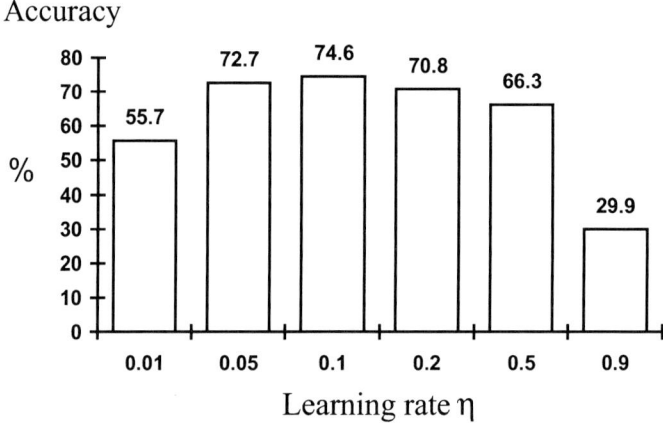

Figure 5.9 Effect of varying the learning rate

From these results, it can be seen that the learning rate is the most critical parameter. Its value cannot be too large otherwise this will cause oscillations during training and the network will not learn. It was found that the momentum did not greatly affect the training except when its value was very high. In this case, the network becomes unstable as it does with relatively large values for the learning rate. For the given classification task, the best value for the momentum was zero, which shows that the momentum was not required at all during training. The number of hidden neurons did not affect performance to a large extent as long as a sufficient number was chosen. In experiments, it was found that equation (5.3) was a good metric for determining the number of hidden neurons.

Accuracy

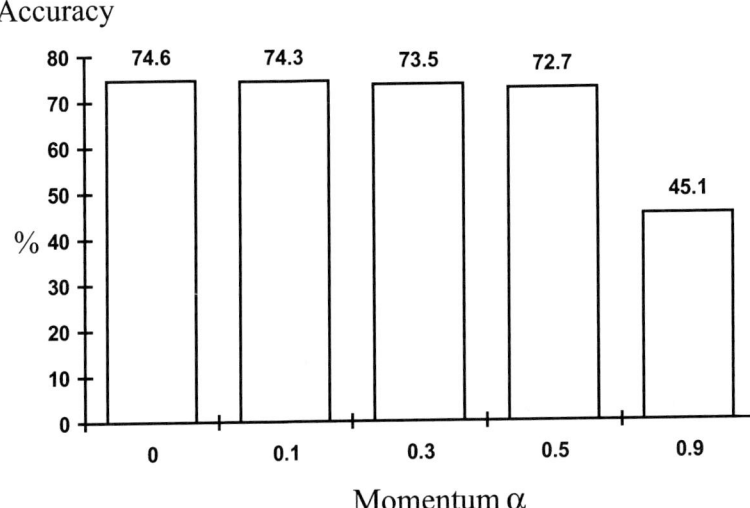

Figure 5.10 Effect of varying the momentum

Accuracy

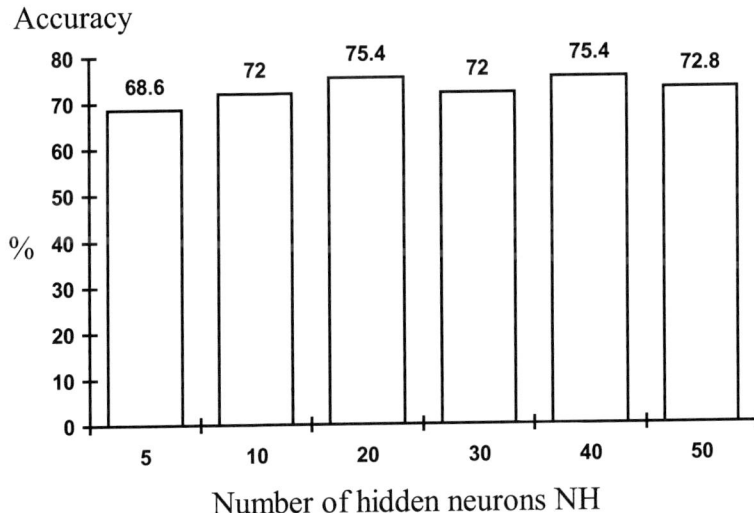

Figure 5.11 Effect of varying the number of hidden neurons

Pham and Sagiroglu [2001] carried out work on classification of objects from AVI images. Different learning algorithms for the MLP were tried to see if they could improve the performance. The tested algorithms were basic BP, delta-bar-delta (DBD), extended DBD and Quickprop. It was found that the standard BP algorithm performed better than the other algorithms and also had the advantage of being simpler.

For the inspection of nuclear fuel pellets, Keyvan et al. [1999] tested a supervised neural network (MLP) against an unsupervised neural network (ART2). Supervised neural networks have access to the class labels for the data whereas unsupervised networks operate without them. Thus, it would be expected that the supervised network would give superior results. This was confirmed by the results obtained.

5.4 Synergistic Classification

Synergy is the effect of combining several units, which exceeds the sum of the individual contributions of the units. In many situations, systems that contain units working together in synergy can produce better results than a single system. For example, in a company, it is often considered preferable to have decisions made by a board than by just one person. Synergistic Classification Systems (SCSs) contain several individual classifiers. The outputs of the classifiers are combined together in some manner to produce a new output. The output of the SCS as a whole depends upon the outputs of the individual classifiers it contains.

SCSs are relatively new and as such there is no standard way to refer to them. They have been called many different names, including Committee Networks [MacKay, 1994], Modular Networks [Haykin, 1994], Hybrid Systems [Goonatilake and Khebbal, 1995], Composite Systems [Pham and Oztemel, 1996] and Multiclassifiers [Mitzias and Mertzios, 1996].

SCSs may contain classifiers of just one type or may combine the benefits of different types of classifier. For example, in the classification of poultry, Chao et al. [1999] combined fuzzy logic and neural networks to create a neuro-fuzzy classifier.

SCSs can be divided into two types, based on how the classifiers are organised. The first is to execute the classifiers in parallel and to have a combination module mediate their outputs. The second is to combine the classifiers into a tree-like structure.

5.4.1 Synergy using a Combination Module

The structure of a SCS that uses a combination module to combine the individual classifiers is given in Figure 5.12. This system contains m classifiers each with n outputs. A combination module, also having n outputs, processes the individual outputs of the classifiers. The largest of the outputs of the combination module is then taken to be the overall output by a Maxnet module.

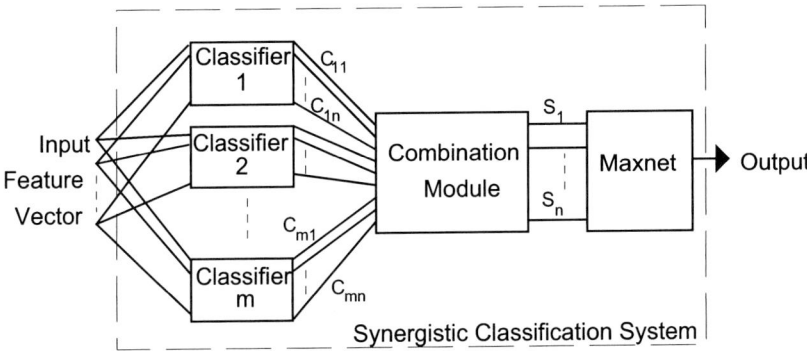

Figure 5.12 Synergistic classification system using a combination module

For a given feature vector, the SCS must combine the outputs of the individual classifiers to provide its own output. This section describes five combination strategies: *summation, maximum, voting, median* and *product*. Another possible combination technique is to use a neural network but this requires additional training and so is not so straightforward to implement. Neural combination modules have been tried previously by several researchers [Bayro-Corrochano, 1993; Sagiroglu, 1994; Mitzias and Mertzios, 1996].

Here, a study of simple combination strategies will be given. Let C_{ij} be the output of classifier i ($1 \leq i \leq m$) for defect type j ($1 \leq j \leq n$) and S_j the output of the combination module for output j. The output of the Maxnet is j if $S_j > S_k$ $\forall k, j \neq k$ ($1 \leq j, k \leq n$). If two or more outputs of the combination module are equal, one possible conflict resolution strategy is for the Maxnet to choose the classifier that gave the best performance on the training set. In other words, the best classifier during training is given the casting vote.

Summation

The corresponding outputs of each classifier are summed and these are then the outputs of the combination module. This technique can be viewed as allowing each classifier to give evidence for every type and then summing all the evidences from each classifier. Thus, output j of the combination module is:

$$S_j = \sum_{i=1}^{i=m} C_{ij} \ (1 \le j \le n)$$

(5.4)

Each S_j can be divided by m to give the average output of the classifiers but this is only a linear scaling operation, which would not alter the final output of the combination module.

Maximum

Output j of the combination module is taken as the maximum output j of all the classifiers, that is:

$$S_j = C_{ij} \mid C_{ij} > C_{kj} \ \forall k, k \ne i \ (1 \le i, k \le m) \ (1 \le j \le n)$$

(5.5)

Voting

Each classifier has one vote as to which defect type it considers the input to be. The combination module then sums the votes. Thus, the Maxnet decides the defect type according to which output has the most votes. This technique is analogous to an election where the candidate with the most votes wins. To calculate which defect is voted for, the maximum output of each classifier is found and this output is changed to a 1. All other outputs of the classifier are set to zero. The new outputs of the classifiers are stored as V_{ij} where:

$$V_{ij} = 1: \text{if } C_{ij} > C_{kj} \ \forall k, k \ne i \ (1 \le i, k \le m) \ (1 \le j \le n)$$
$$V_{ij} = 0: \text{otherwise}$$

Thus, $V_{ij} = 1$ if classifier i votes for defect type j

The output of the combination module is:

$$S_j = \sum_{i=1}^{i=m} V_{ij} \ (1 \le j \le n)$$

(5.6)

Median

The median combination strategy uses the middle value of the classifier outputs for each type after they have been sorted into numerical order. The potential advantage of the median is that it suppresses the highest and lowest outputs, which could be outliers. To calculate the median, first the outputs of the classifiers are sorted. So, for each defect type j, an ordered version O_{ij} of C_{ij} is created where O_{1j} is the smallest C_{ij} ($1 \leq i \leq m$) and O_{mj} is the largest. Then, the output of the combination module is:

$$S_j = O_{kj} \quad (1 \leq j \leq n)$$

(5.7)

where $k = (m+1)/2$, assuming m is an odd integer. If m is an even integer then:

$$S_j = \tfrac{1}{2}(O_{kj} + O_{k+1\,j})$$

(5.8)

where $k = m/2$.

Product

This strategy is similar to the summation strategy but instead of adding the outputs, they are multiplied:

$$S_j = \prod_{i=1}^{i=m} C_{ij} \quad (1 \leq j \leq n)$$

(5.9)

As the outputs of the MLP are between zero and one, the product strategy gives smaller results than summation when the classifiers do not agree. For example, with a two-classifier system, if one classifier outputs 0.5 and the other classifier gives 0.5 then the summation output will be 1.0 and the product output will be 0.25. However, if one classifier produces 0.1 and the other 0.9, then the summation output will still be 1.0 but the product will be 0.09.

Experiments were carried out using the different combination strategies and the results are shown in Table 5.1. Thirty MLP classifiers were trained. For a three-classifier SCS, the best three classifiers were selected.

		Accuracy	%				
		Sum	Max	Voting	Median	Product	Average
	3	92.0	92.0	92.0	92.0	92.0	92.0
	5	92.0	92.0	92.0	92.0	92.0	92.0
C	7	92.0	90.9	90.9	92.0	92.0	91.6
	9	90.9	89.8	90.9	92.0	89.8	90.7
	11	89.8	88.6	90.9	90.9	89.8	90.0
	Av.	91.3	90.9	91.3	91.8	91.1	-

Sum – summation   Max – maximum   C – number of classifiers

Table 5.1 Performances of the synergistic classification systems

The following were concluded from the experiments:

1. Using three or five classifiers gave the best performance of 92.0% irrespective of the combination strategy chosen.   Since three classifiers require less computation than five and produce the same results, three was considered to be the optimum number of classifiers to use in the SCS. The performance of the best SCS was 1.1% better than that of the best individual classifier (90.9%). In AVI, this improvement in accuracy can be considered sufficient to warrant the extra computation required by the use of more than one classifier and the combination strategy.

2. Employing more than five classifiers gives progressively worse results.   One explanation for this is that the average performance of the best seven classifiers is lower than the average performance of the best three classifiers.

3. When using three or five classifiers, all combination strategies performed equally well.   However, when average performances are taken for differing numbers of classifiers, some strategies produced slightly better results. The best strategy overall was the median strategy which gave an average accuracy of 91.8%. The maximum strategy yielded the worst performance with a mean of 90.9%. However, the median strategy requires a sorting operation and so will take slightly longer to execute than the other methods.

Hypothetically, in a SCS, if each of the individual neural networks classifiers were trained using different parameter values then this should result in more diverse component systems than if they were all trained with the same parameter values. These diverse component systems can then be combined to create a SCS having a wider coverage of the feature space.   To test this hypothesis, ten networks were trained with different parameter values from one another and ten networks were trained utilising the same parameter values. To obtain parameter values for the first set of networks, parameter values randomised around the optimum values were calculated. Ten networks were trained of each type and the best three selected in both cases. It was found that the SCS consisting of classifiers trained using the same parameters performed better, thus refuting the above hypothesis.

In the work of Mitzias and Mertzios [1996], three MLP classifiers were trained, with each classifier having a different set of features as input. Therefore, each classifier gains expertise in a specific area of the domain. The first classifier employed geometric features. The second utilised 1-D scaled normalised central moments. The third classifier used angles of a fast polygon approximation method. The best performance was obtained when the outputs of all three classifiers were combined together.

Oh et al. [1999] also worked on synergistic classifiers with different inputs. The idea of the work is illustrated in Figure 5.13. This figure shows a simple feature space involving three classes, labelled with squares, circles and triangles. It can be seen that squares can be classified just using feature 1. Feature 2 is needed to differentiate between circles and triangles.

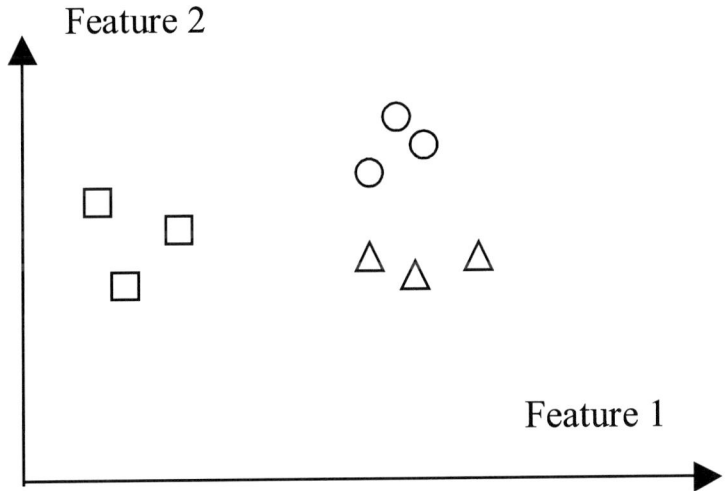

Figure 5.13 Feature space containing three classes

Figure 5.14 shows a synergistic classification structure suitable for this problem. Classifier 1 utilises feature 1 to identify the square class, whilst classifier 2 concentrates on differentiating between circles and triangles using feature 2.

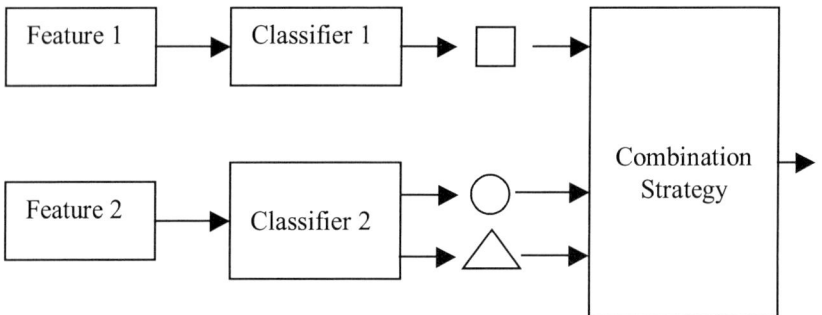

Figure 5.14 Synergistic classifiers with different inputs

5.4.2 Tree-based Synergistic Classifiers

Drake and Packianather [1998] developed tree-based SCSs. Instead of integrating the outputs of several neural networks using a combination module, they connected them in a tree structure. To classify a given feature vector, first, the vector is passed to an individual MLP at the top of the tree. If the MLP determines that the type of the feature vector is one that it classified with 100% accuracy in training, then that type is accepted. Otherwise, the features are passed to a second level classifier. Each second-level classifier is designed to differentiate between just two types. Therefore, second-level classifiers are able to concentrate on a smaller area of the feature space.

Figure 5.15 shows a simple illustrative example. Here, a feature vector is to be classified into one of three types. The features are presented to a neural network called NN1. If the network determines that the features are of class 2, which is the class it was able to recognise perfectly during training, then classification is terminated. Otherwise, the features are fed to a second network (NN2) that is specialised in discriminating between classes 1 and 3.

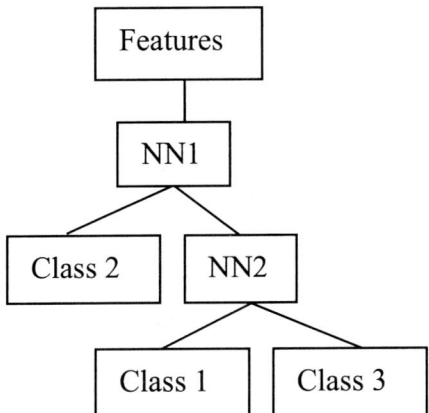

Figure 5.15 Tree-based synergistic classification

This technique was subsequently improved such that features are selected specifically for each second-level classifier. Therefore, each network uses only the features that it requires and so is specialised in a certain portion of the feature space.

5.5 Discussion

When selecting a classifier for AVI, research that has compared the performances of the different types of classifier is relevant. Koivo and Kim [1986] measured the performance of a tree classifier, a minimum-distance classifier and a Bayesian classifier, finding the tree classifier to be the best. Tree classifiers are similar to rule-based classifiers and can be implemented using rules. Cho and Conners [1991] and Cho et al. [1991] compared the performance of a MLP, a KNN statistical classifier and a rule base. Both studies showed that the neural network gave the best performance. It was also stated that the neural network was easier to implement than the rule-based system.

Brzakovic and Vujovic [1996] compared different classifiers for web inspection. The three classifiers that were tried were the MLP, the SOFM and the Bayesian classifier. It was found that all classifiers gave an accuracy of around 85%, so the classifier choice was not of great importance.

A useful comparison of classifiers comes from the STATLOG project, a European ESPRIT initiative to determine the effectiveness of different classifiers on a wide-range of data sets [Michie et al., 1994]. The classifiers tested were statistical algorithms, inductive decision trees, automatically induced rule sets and neural networks. The statistical algorithms included discriminant functions, nearest-neighbour, KNN and Bayesian classifiers. The nearest-neighbour classifier learned slowly on very large data sets. Sampling the training data set can reduce this problem. The Bayesian classifier gave better results when the features were independent given the class. The decision tree classifiers tested were similar to the classic ID3 algorithm and included NewID, C4.5 and CART. All decision tree classifiers gave a similar performance. Decision tree methods gave an advantage over statistical methods when there were a large number of features. CN2 was tested as an example of induction methods of producing rule sets for classification. The performance of CN2 was found to be very similar to those of decision trees. The neural networks tested included the MLP, SOFM, LVQ and RBF networks. It was found that neural networks gave the best or near-best performance in nearly every case. One problem that was identified was the difficulty in choosing optimal network parameter values. However, LVQ was found to be simple to use and executed quickly.

A further consideration in classification is the cost of misclassifications. In industries where quality is very important, such as the aeronautical industry, it may be preferable to reject many good samples, so that no bad samples are allowed to pass. Rejected components can then be sent for detailed manual inspection. Neural networks do not inherently have this capability. However, Bayes theory has been extended to cover the cost of misclassifications and CART can also incorporate costs into its classification.

The MLP classifier is often chosen to be the classifier in AVI systems. This is because it gives a good performance and does not require a large effort to train. To obtain optimal performance with the MLP, or to improve its performance, several factors should be taken into account. First, the learning rate should be set to a relatively low value to avoid oscillations during training. Second, the size of the feature vector used as input to the MLP should be sufficiently large. Third, more than one MLP can be employed to obtain a more robust classifier.

5.6 Summary

Classification is the stage of AVI that identifies the objects found during the segmentation stage. The result of the classification is either the grade of the object being inspected or a decision as to whether the object should be accepted or rejected.

Many different types of classifiers have been employed for AVI tasks. These include statistical, rule-based and neural network classifiers. The MLP neural network is both simple to use and effective in its operation. Recent research attempts have been directed at combining several classifiers together into Synergistic Classification Systems. Such systems are more robust than individual classifiers.

References

Bayro-Corrochano E.J. (1993) *Artificial Intelligence Techniques for Machine Vision*. PhD thesis, School of Engineering, Cardiff University, UK.

Brzakovic D. and Vujovic N. (1996) Designing a Defect Classification System: A Case Study. *Pattern Recognition*. Vol. 29, No. 8, pp. 1401 - 1419.

Chao K., Chen Y.R., Early H. and Park B. (1999) Color Image Classification Systems for Poultry Viscera Inspection. *Applied Engineering in Agriculture*. Vol. 15, No. 4, pp. 363 - 369.

Cho T.H. and Conners R.W. (1991) A Neural Network Approach to Machine Vision Systems for Automated Industrial Inspection. *Int. Joint Conf. on Neural Networks*. Seattle, WA. pp. I 205 - 211.

Cho T.H., Conners R.W. and Araman P.A. (1991) A Comparison of Rule-Based, K-Nearest Neighbour and Neural Net Classifiers for Automated Industrial Inspection. *Proc. IEEE / ACM Int. Conf. on Developing and Managing Expert System Programs*. Washington, DC. pp. 202 - 209.

Davies E.R. (1996) *Machine Vision: Theory, Algorithms, Practicalities*. (2nd ed.) Academic Press, London.

Drake P.R. and Packianather M.S. (1998) A Decision Tree of Neural Networks for Classifying Images of Wood Veneer. *Int. Journal of Advanced Manufacturing Technology*. Vol. 14, pp. 280 - 285.

DTI (1994) *Best Practice Guidelines for Developing Neural Computing Applications*. Electronics and Engineering Division, Department of Trade and Industry, 151 Buckingham Palace Road, London, UK.

Goonatilake S. and Khebbal S. (1995) *Intelligent Hybrid Systems*. John Wiley and Sons, Chichester, UK.

Haykin S. (1994) *Neural Networks: A Comprehensive Foundation*. Macmillan, New York.

Keyvan S., Song X.L. and Kelly M. (1999) Nuclear Fuel Pellet Inspection using Artificial Neural Networks. *Journal of Nuclear Materials*. Vol. 264, No. 1-2, pp. 141 - 154.

Koivo A.J. and Kim C.W. (1986) Classification of Surface Defects on Wood Boards. *IEEE Int. Conf. on Systems, Man and Cybernetics*. Atlanta, GA. pp. 1431 - 1436.

MacKay D.J.C. (1994)   Bayesian Non-Linear Modelling for the Prediction Competition. *ASHRAE Transactions*. Vol. 100, Pt. 2, ASHRAE, Atlanta, GA. pp. 1053 - 1062.

Magee M., Weniger R. and Wenzel D. (1993)   Multidimensional Pattern Classification of Bottles using Diffuse and Specular Illumination. *Pattern Recognition*. Vol. 26, No. 11, pp. 1639 - 1654.

Michie D., Spiegelhalter D.J. and Taylor C.C. (1994) *Machine Learning, Neural and Statistical Classification*. Ellis Horwood, New York and London.

Mitzias D.A. and Mertzios B.G. (1996) A High Performance System for Generic Pattern Recognition Applications. *Proc. 3rd Int. Workshop on Image and Signal Processing*. Manchester, UK. pp. 357 - 360.

Oh I.S., Lee J.S. and Suen C.Y. (1999) Analysis of Class Separation and Combination of Class-Dependent Features for Handwriting Recognition. *IEEE Trans. on Pattern Analysis and Machine Intelligence*. Vol. 21, No. 10, pp. 1089 - 1094.

Pham D.T. and Alcock R.J. (1999a) Synergistic Classification Systems for Wood Defect Identification. *Proc. IMechE. Part E - Journal of Process Mechanical Engineering*. Vol. 213, No. 2, pp. 127-133.

Pham D.T. and Alcock R.J. (1999b) Automated Visual Inspection of Wood Boards: Selection of Features for Defect Classification by a Neural Network. *Proc. IMechE. Part E - Journal of Process Mechanical Engineering*. Vol. 213, No. 4, pp. 231 - 245.

Pham D.T. and Dimov S.S. (1997) An Efficient Algorithm for Automatic Knowledge Acquisition. *Pattern Recognition*. Vol. 30, No. 7, pp. 1137 - 1143.

Pham D.T. and Oztemel E. (1996) *Intelligent Quality Systems*. Springer-Verlag, Berlin and London.

Pham D.T. and Sagiroglu S. (2001) Training Multilayered Perceptrons for Pattern Recognition: A Comparative Study of Four Training Algorithms. *Int. Journal of Machine Tools and Manufacture*. Vol. 41, No. 3, pp. 419 - 430.

Ravi V. and Zimmermann H.J. (2000) Fuzzy Rule Based Classification with FeatureSelector and Modified Threshold Accepting. *European Journal of Operational Research*. Vol. 123, No. 1, pp. 16 - 28.

Sagiroglu S. (1994) *Modelling a Robot Sensor using Artificial Neural Networks*. PhD thesis, School of Engineering, Cardiff University, UK.

Shen Q. and Chouchoulas A.A. (2000) Modular Approach to Generating Fuzzy Rules with Reduced Attributes for the Monitoring of Complex Systems. *Engineering Applications of Artificial Intelligence*. Vol. 13, No. 3, pp. 263 - 278.

Webb A. (1999) *Statistical Pattern Recognition*. Arnold, London.

Problems

1. Classification experiments are performed for nine objects of three types: A, B and C. The following table shows the results of classification. The type that the object should be classified into is designated by *Real Type* and the actual type output by the classifier is shown in the column *Classified Type*.

Object	Real Type	Classified Type
1	A	A
2	A	A
3	A	A
4	B	B
5	B	C
6	B	B
7	C	B
8	C	B
9	C	B

Draw the confusion matrix for the given results.

2. The following training data shows the values of two features and the corresponding output class:

Feature 1	Feature 2	Class
1	2	A
1	3	A
2	2	A
2	3	A
2	4	B
2	5	B
3	4	B
3	5	B

Write if-then rules that could be used to classify this training data.

3. Using minimum-distance classification, determine the class of an object which has feature 1 equal to 3 and feature 2 equal to 3.5.

4. Would employing the KNN classifier (with K=3) give a different result from the minimum distance classifier in this case?

5. Given the following test data, determine the accuracy of the rule(s) generated in Problem 2.

Feature 1	Feature 2	Class
3	1	A
3	2	A
3	3	A
4	3	B
4	4	B
4	5	B

If the accuracy of the generated rules is less than 100%, give a reason why this might be so and a suggestion for improving results in the future.

6. In a synergistic classification system with three classifiers, the following outputs are generated for three types:

	Type 1	Type 2	Type 3
Classifier 1	0.7	0.9	0.9
Classifier 2	0.7	0.7	0.8
Classifier 3	0.7	0.4	0.3

Determine the output of the whole system, using the summation, maximum and median combination strategies.

7. In ProVision, the *M Decision Making* function is employed for classification. Read the section "M Decision Making Dialog Box" in the online manual. Describe how classification is carried out in ProVision.

8. Open the DATE project. Run the project, putting a breakpoint at the end of the cycle. What is the output of the bit pattern? In the decision list, remove the +H in column C3 of serial number 1. What is the new output of the bit pattern? What contributions do columns C1, C2 and C3 give to the overall output?

Chapter 6

Smart Vision Applications

This chapter describes examples of smart-vision applications in automotive component manufacture, veneer board inspection and textile production.

6.1 Inspection of Car Engine Seals

Valve-stem seals are an important engine component. Their role is to prevent excessive oil from running down the valve stem and into the cylinder, where it will be burnt causing pollution. Defects occur in the production of valve-stem seals mainly during injection moulding.

Car engine seals can be produced at a rate of one per second or even faster. At these speeds, 100% manual inspection is not possible. Therefore, the approach to manual inspection is only to inspect samples of seals (around 1% to 2% of all seals produced). However, it is very important to have all seals inspected, particularly when selling to manufacturers of high quality cars. Another problem is that the inspection is off-line, so by the time defects are found, a large number of defective seals have already been produced.

To design an AVI system for seals, three types of knowledge are required. First, the types of seal defects and their locations must be known. Second, heuristics about defects' sizes and shapes are required. Finally, it is important to obtain knowledge on the effect of each defect on the functionality of the seal.

To obtain the knowledge necessary, quality control personnel were interviewed. Typical examples of defective seals were collected to illustrate the possible flaws. In total, nine defect types were found: blister, lobed, spring lip damage, contamination, flow mark, old rubber, high injection gate, cut oil angle and scorch.

6.1.1 System Set-up

It was decided that one camera would be insufficient to locate all the important defects. Thus, the arrangement chosen was one top camera and three angled cameras placed around the seal at a spacing of 120° [Pham et al., 1995]. Each of the three cameras viewed one third of the top of the seal with slight overlaps. The arrangement is shown in Figures 6.1 and 6.2.

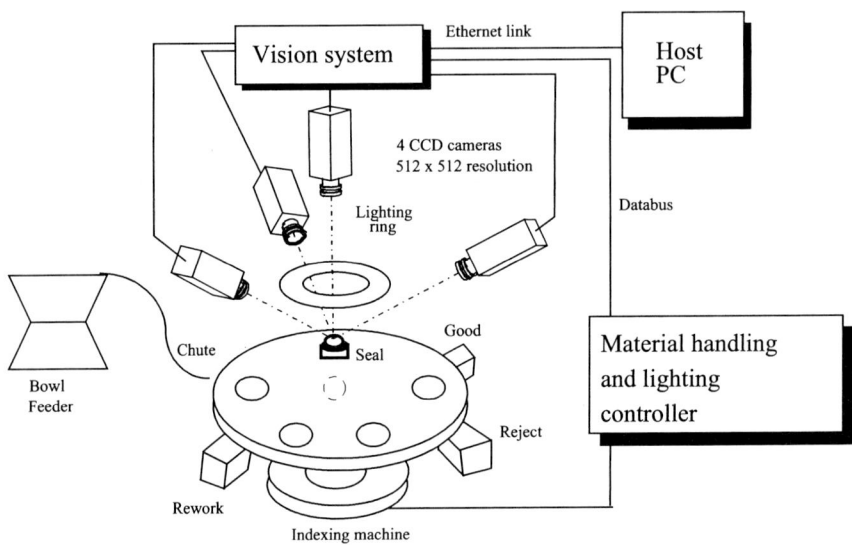

Figure 6.1 Inspection system for car engine seals

As in all inspection tasks, it is essential that an optimum lighting arrangement be designed. For lighting from behind the seal, a diffused back light was found to give good results. Motivation for this also came from the information that human inspectors find this lighting method to be good for seal inspection. To create the back light, four LEDs were placed below a Perspex diffuser.

For the front lighting, there was the requirement that the lights needed to be switched on and off quickly so that only the back light showed for a given duration during inspection. Filament lamps could not meet this requirement and LED lighting was chosen. Strobe lighting would have required the use of an expensive "charge inhibit" camera. Other advantages of LED lighting include long life, constant illumination level, low power requirements and small physical size. Trial-

and-error experiments were carried out to determine the optimal angle to place the lighting. This was found to be 20° to the tangent of the seal surface. Sixteen LEDs were mounted in a circular aluminium holder and angled at 20° to create the required front light.

Figure 6.2 Seal inspection system

For the mechanics of the inspection system, the seals are fed into a bowl feeder that places them one by one onto a rotating index table. The table moves the seals in turn into position to be inspected. The LED lighting ring then illuminates the seals from above and, next, the backlight shines through the hole in the seals. Four cameras are used, one from above the seal to obtain an image of the inside of the seal and three from the sides to inspect the lip edge. The inspection is performed on a dedicated inspection system, which is controlled by a host PC acting as a dumb terminal [Performance Vision, 1992]. After analysis, the seals are placed into one of three bins, according to the decision made by the inspection system. The bins are labelled "accept", "reject" and "rework".

6.1.2 Inspection Process

The sealing lip of the seal should be a perfect circle when placed on the valve stem for optimum performance. However, due to natural flexing of the rubber, the seal is not perfectly circular during the inspection process. An algorithm was developed to detect defects whilst being insensitive to allowable "out of circularity". The algorithm is described below:

1. Switch on the backlight only to acquire an image of the sealing lip contour with the top camera. The camera resolution was chosen so that defects smaller than two pixels would be ignored.

2. Threshold to produce a binary image. Due to the lighting adopted, the image was almost binary already, so a fixed threshold could be used. The result of this operation is a white object on a black background.

3. Obtain the Freeman [1974] chain code of the object to represent the object's position in row and column co-ordinates.

4. Determine the co-ordinates of the object centre.

5. Correct the shape of the image by removing the optical distortion introduced by the non-unity aspect ratio of the camera. This step would produce a perfect circle if the sealing lip is defect free.

6. Calculate the radial distance (r) of each point on the edge of the object.

7. Subtract the average of fifty consecutive r values from the central (25th) r value. This is repeated for all r values. This step removes the slight radius variation caused by the natural flexing of the rubber.

8. Use an averaging filter to reduce the effect of quantisation noise. The filter operates on five data points at a time and has a triangular weighting arrangement with the maximum weight at the central point.

9. Apply high and low thresholds to the data. A point exceeding these thresholds indicates a significant deviation from the nominal radius and is thus a point of interest (POI).

10. Classify the seal as good if the number of POIs is less than three. Note that there are around 800 pixels on the seal contour.

11. If there are at least three POIs then the seal is rejected. Three features are extracted: the width of the defect (W), the internal deviation from the nominal radius (INTDEV) and the external deviation from the nominal radius (EXTDEV). Note that all measurements are in pixels.

12. A rule-based expert system is employed to classify the type of defect. The rules developed are as follows:

- if W>300 then seal is lobed;
- if W>50 and EXTDEV<4 and EXTDEV>INTDEV then seal is contaminated;
- if W>50 and EXTDEV>4 and INTDEV>4 then seal is scorched;
- if W>50 and EXTDEV>4 then seal contains old rubber;
- if W<50 then seal is blistered.

Experiments were carried out using 100 good seals and 100 seals with defects of various types. These seals were used repeatedly to simulate a working production system. Only 1 in 100,000 faulty seals were falsely classified as being good. Conversely, just 1% of good seals was incorrectly classified as being defective.

For the classification of the defective seals, the results are shown in Table 6.1. It can be seen that the average performance of the system was 67%, with the classification of old rubber being particularly poor at 33%. The relatively low performance was attributed to the simple nature of the features used for reasons of speed.

Defect type	Performance
Lobed	90%
Blister	68%
Contamination	64%
Scorched	48%
Old rubber	33%
Overall	**67%**

Table 6.1 Performance of rule-based defect classifier

Pham and Bayro-Corrochano [1994] proposed different features from those used above. After the image had been thresholded (step 2), the Laplacian filter is employed to obtain the edge contour of the seal. Then, twenty convolution masks are applied to obtain measures of the shape of the circular centre of the seal. The twenty masks are shown in Figure 6.3.

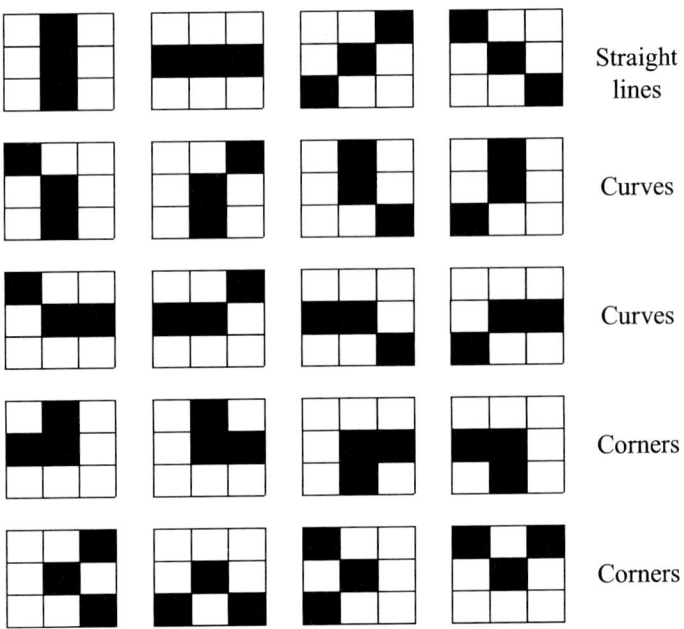

Straight lines

Curves

Curves

Corners

Corners

Figure 6.3 Set of geometric features

Figure 6.4 shows three defect types and histograms of the values of the twenty geometric features. The histograms show the number of times that each of the masks matches the edge contour of the seal.

In addition to these twenty features, five other geometric features were adopted. These were the area of the attribute, its perimeter, the length and width of its minimum bounding rectangle and the area of the latter.

Two training sets, each of 180 feature vectors were used. Two neural networks were tested for classification, the back-propagation multi-layer Perceptron (BPMLP) and LVQ networks. The BPMLP took 150,000 iterations to get its global output error below 0.01. The LVQ network required 3,600 iterations to reach its minimum global error of 0.02. Two test sets, each with 100 feature vectors were employed. The BPMLP was able to classify correctly 83% of these vectors. The LVQ network only achieved 74%.

To improve upon this performance, Pham and Bayro-Corrochano [1995] tested other neural network architectures. Again, the twenty geometrical features were employed but this time, only three other attributes were added, giving feature vectors of dimension 23. The networks considered were the adaptive logic network (ALN), the BPMLP and the Kohonen SOFM. The ALN, developed by Armstrong and Gecsei [1979], is a feedforward network where the nodes only compute Boolean functions. The nodes are connected in a tree-like structure. The ALN was able to identify correctly 78% of the test patterns.

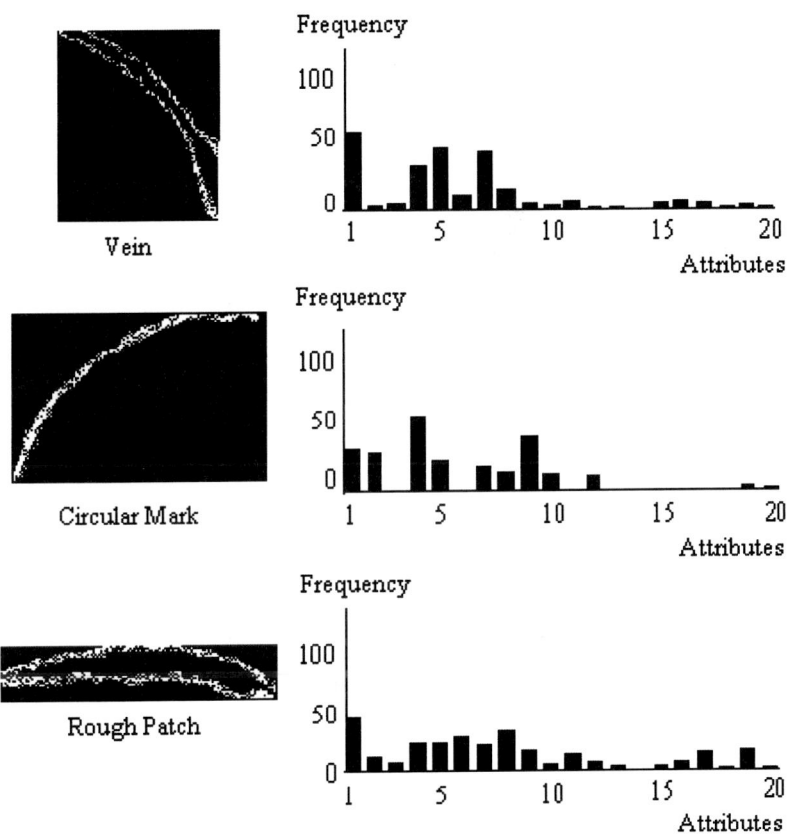

Figure 6.4 Defects and corresponding geometric feature histograms

For the BPMLP, a three-layer network was chosen with twenty-three input neurons, ten hidden neurons and three output neurons. A learning rate of 0.7 and a momentum of 0.8 were chosen. The network was trained for 1.5 million iterations to reduce the global error to below 0.001. The resulting BPMLP classified 90% of the feature vectors correctly.

The Kohonen SOFM consisted of a square grid of 10x10 neurons. A square neighbourhood was adopted during training, a process that involved almost 18,000 iterations. The resulting labelled map classified 79% of vectors.

To increase performance further, synergistic classification was considered. It was observed that the defect types could be divided into their sizes, corresponding to small, medium and large. Classification of all three types was difficult using just one network. Therefore, a synergistic solution was proposed with three networks, one for each of the defect sizes. Three BPMLPs were used and a fourth BPMLP as the combination module. BPMLPs were chosen because of their superior individual performances compared to the other networks. Another advantage of BPMLPs is that they give continuous outputs, which can be employed as confidence values for their decisions. These confidence values are useful in the combination module.

The combination module BPMLP had twelve input neurons, three from each of the three networks as well as the other three attributes. It required approximately 300,000 iterations to reach a global output error of 0.01. The resulting synergistic system was able to classify 93% of defects correctly.

6.1.3 Subsequent Work

The above-mentioned seal inspection project was completed in 1994 and produced a working system enabling the seal manufacturer to improve productivity, reduce scrap and effect large process savings. Upon completion of the project, a follow-up project was instigated. The idea of the second project was to pass information back from the grading to the production process. For example, if many examples of a certain type of defect are found, then this may indicate a particular fault with one of the production machines.

Pham and Bayro-Corrochano [1995] describe such a quality improvement system (Figure 6.5). The system consists of an AVI machine, a process monitoring computer (PMC) and a quality management computer (QMC). Objects are supplied to the AVI machine for inspection. This provides information on the quality of the objects to the QMC. Simultaneously, diagnostic information is sent to the QMC by the PMC, which monitors selected parameters from the production process. These parameters could include factors such as machine temperature, pressure and cycle time. The PMC contains an expert system for statistical process control and diagnosis [Pham and Oztemel, 1996]. Using a model of the manufacturing process, which relates to the observed process conditions and the process parameter settings, the QMC computes and transmits feedback information to the production machines to adjust them.

In related work on the project, Pham and Peat [1995] worked on analysis of the seal manufacturing process using a synergistic approach.   The three sources of knowledge used were a system model, human experience and recorded data from the system.   The framework for the method was the system model, which was stored in the form of a directed graph.   Nodes represented processes and arcs showed information flow.   An important part of the method was the highly visual, automatic and intuitive approach.

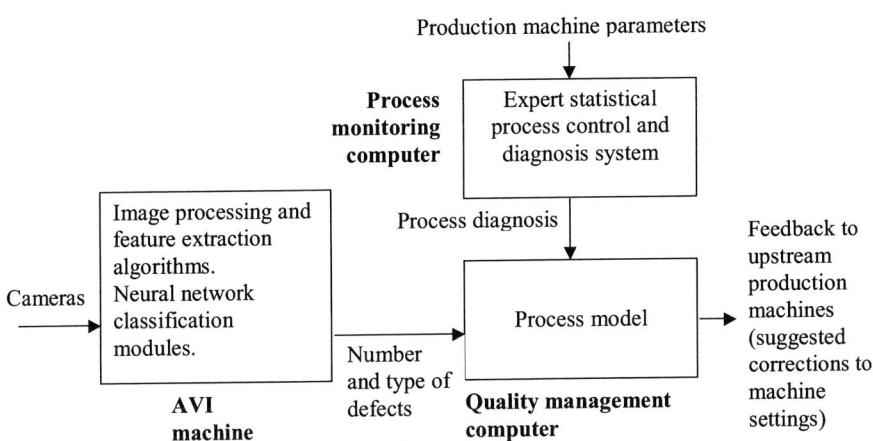

Figure 6.5 Concept of quality improvement system

6.2 Inspection of Wood Boards

The production of veneer boards involves bonding together thin sheets of wood. Reviews of AVI research in this area can be found in [Pham and Alcock, 1998a, b]. Normally, grading of these sheets is carried out to ensure that only high quality sheets are used to make high quality boards.   The number, size and type of defects on the sheet surface determine its grade.   Currently, in many veneer factories, a human inspector performs this grading.   However, two problems can occur with manual grading, which are called *downgrading* and *upgrading*.   Downgrading happens when a sheet is placed into a grade lower than its correct grade.   This causes value to be lost because a lower grade veneer can only be sold for a lower price.

Conversely, upgrading arises when a sheet is placed into a grade higher than it should be put into. When this happens, the sheet must be manually re-graded, requiring time and effort because the sheet may need to be moved to a different part of the factory. In either case, an error in grading causes a loss in revenue. These errors occur because the grader typically has only one second in which to grade the sheet and becomes mentally fatigued by the monotony of the job. Experiments have been performed that confirm the accuracy of human wood graders to be well under 100%. Huber et al. [1985] tested 6 willing rough mill employees and found that their accuracy was 68%. Polzleitner and Schwingshakl [1992] carried out four independent trials on human graders and observed an average performance of just 55%.

For birch wood, commonly used as the ply material, the defects to be found include coloured streaks, hard rot, holes, pin knots, rotten knots, sound knots, splits, streaks and worm holes. Examples of these defects, acquired using front and back illumination are illustrated in Figure 6.6, together with an example of clear wood. Note that the hard rot example comes from the centre of a large area of hard rot and so the example completely fills the frame.

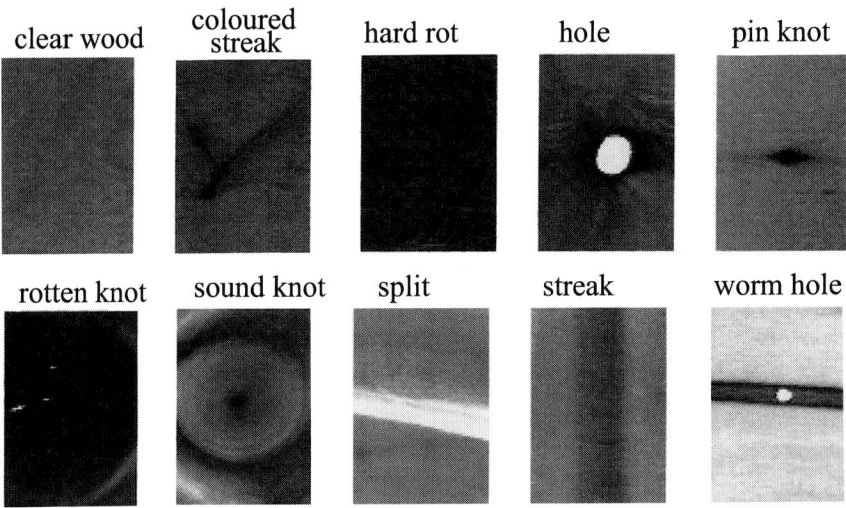

Figure 6.6 Wood veneer defects

6.2.1 System Set-up

An automated inspection system for birch wood veneer boards was designed for the Finnish wood industry as part of a European Union research project. The testbed for the system was set up at the Finnish Wood Research Institute (VTT), in Kuopio, Finland (Figure 6.7).   Figure 6.8 shows the production process of wood veneer boards.   First, a log is soaked, the bark is removed and it is cut into poles of around 1.5m in length.   Next, the poles are peeled, turning them into long thin wood sheets, which are then clipped into square sheets.   Finally, the sheets are dried, graded and bonded together to make veneer boards of different quality grades.

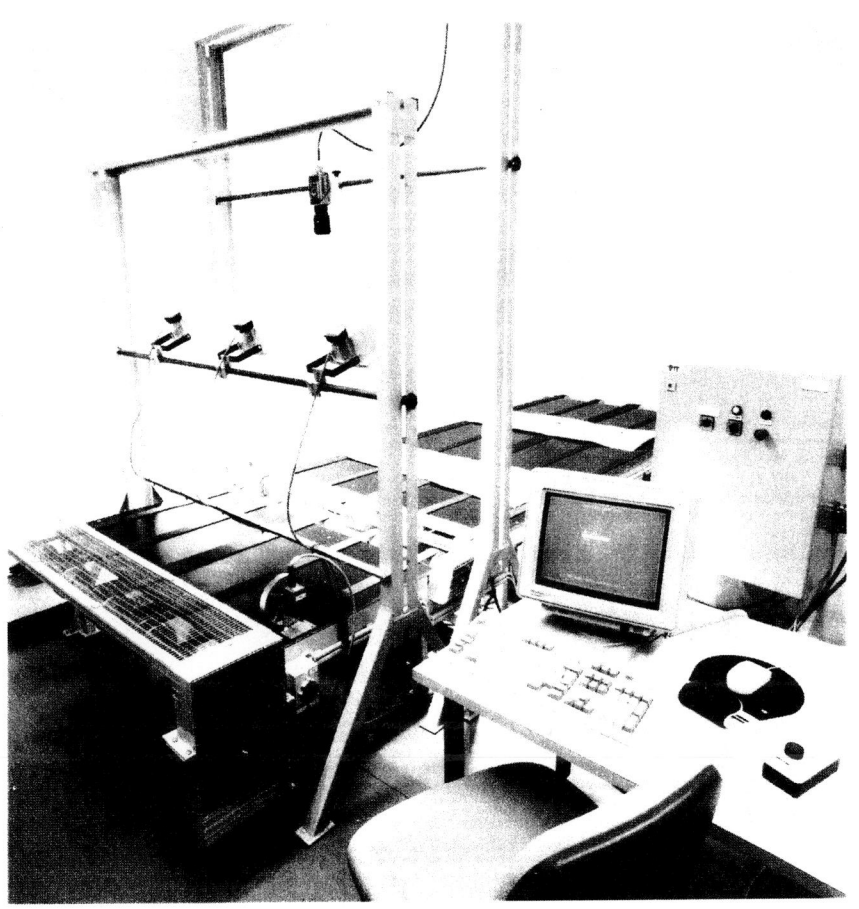

Figure 6.7 Laboratory testbed (image courtesy of Timo Lappalainen, VTT, Finland)

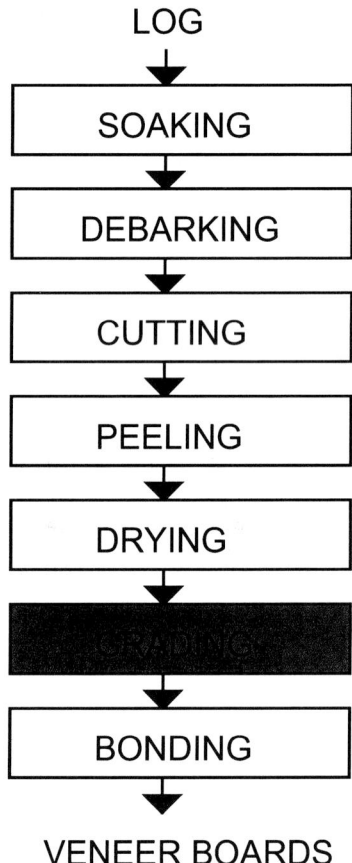

Figure 6.8 Veneer board production process

6.2.2 Inspection Process

The inspection routine developed is illustrated in Figure 6.9. First, an image of the board is acquired using appropriate lighting. Second, defective areas of the board are found using segmentation modules, each specialised in detecting a different type of defect [Pham and Alcock, 1996; 1999a]. Third, post processing is performed to remove false objects and combine areas that represents the same defect [Pham and Alcock, 1998c]. Fourth, both object features and window features are extracted from each located area [Pham and Alcock, 1999b]. Fifth, the features are passed to three neural networks and the outputs of these networks are then combined using the median combination strategy to assign an overall class to each region [Pham and Alcock, 1999c]. Finally, the board is graded, using a set of rules, based on the defect type of each located region. An example of a grading table is shown in Table 6.2. The table shows for each grade, the number, type and size of defects that are permissible. For example, for grade A, only pin knots and a few coloured streaks are allowed. This table can be easily converted into a rule-based expert system. For better results, fuzzy rules can be employed to emulate expert human graders more closely.

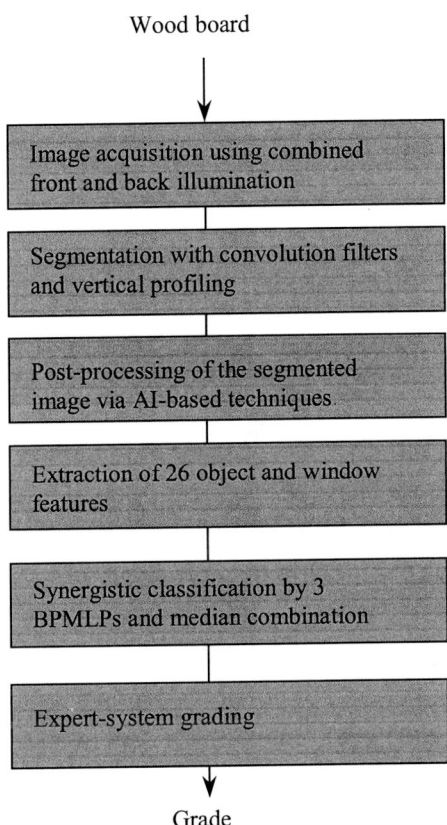

Figure 6.9 Developed system for AVI of wood boards

The segmentation method adopted is based on standard image-processing functions and consists of four stages. Before segmentation, an image of the veneer sheet being inspected is acquired using lighting from above and beneath the sheet. This image is called image A. An example of an acquired image is shown in Figure 6.10(a). This image contains several defects caused by bark. The white areas are where the bark has become too thin and the sheet is perforated.

Defect	Grade A	Grade B	Grade C	Grade D	Grade E
Pin knot	√	√	√	√	√
Sound knot	x	ϕ < 5mm	ϕ < 15mm	ϕ < 25mm	√
Rotten knot	x	x	X	ϕ < 5mm	ϕ < 20mm
Hole	x	x	ϕ < 10mm	ϕ < 10mm	ϕ < 20mm
Split	x	x	l < 100mm	l <200mm	√
Colour streak	Few	< 10% of sheet	√	√	√
Hard rot	x	x	X	x	x
Streak	x	x	√	√	√

√ - permitted   x - not permitted ϕ - diameter   l - length

Table 6.2 Example of a grading table

The four stages of the method are:

1. Image B is created as a copy of image A except that very high and very low grey levels are removed. This is because pixels in an image of a wood board that do not represent defects normally take intermediate grey levels. As mentioned earlier, severe defects such as rotten knots have very low grey levels. When back illumination is employed, holes and splits are very bright. In experiments, all images initially had 256 grey levels. Pixels with a grey level of less than 100 were given a grey level of 100 and pixels with a grey level of more than 180 assigned a grey level of 180. This operation was achieved using a look-up table. The result of the stage, on the image in Figure 6.10(a), is depicted in Figure 6.10(b).

2. Image B is smoothed using a 7x7 averaging convolution filter, resulting in image C which is called the expected image. Figure 6.10(c) shows the outcome of performing 7x7 averaging on the image in Figure 6.10(b).

3. The difference between images A and C is calculated to generate a new image D. Figure 6.10(d) displays image D for the example image. Image D, which has been negated for display purposes, represents the difference between the original and expected images. In Figure 6.10(d), dark pixels correspond to a large variance between the two images. To calculate the difference between two images requires three addition and subtraction operations if images cannot store negative pixel values. First, image B is subtracted from image A and stored in buffer X.

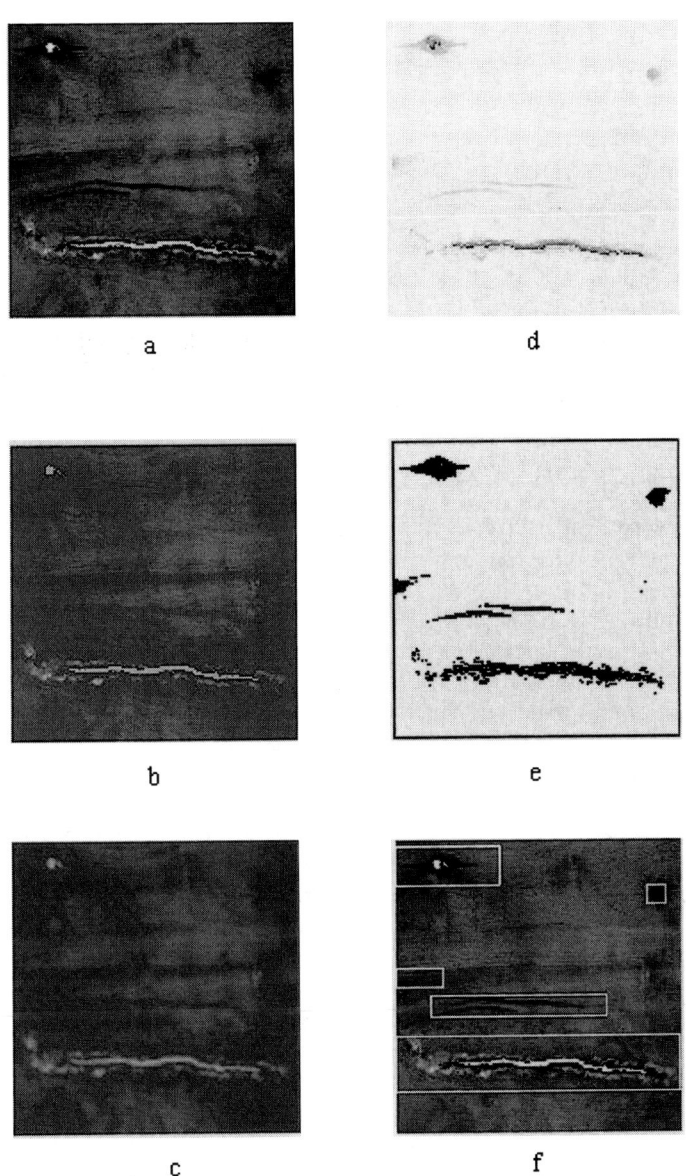

Figure 6.10 (a) Image A - a veneer sheet (b) Image B - image A with very low and high grey levels removed (c) Image C - image B after 7x7 averaging (d) Image D - difference between images A and C (e) Image E - objects found for image A (f) Image A with all defects localised.

Any negative values generated by this operation are stored as zero values. Next, image A is subtracted from image B and stored in buffer Y. Finally, the contents of buffers X and Y are summed to give image D.

4. All pixels in image D with a grey level below a threshold (empirically chosen as 20) are given a grey level of zero. Image E, Figure 6.10(e), is the result of this operation followed by a thresholding operation using a threshold of 1. Only pixels in images A and C that differ by a significant amount are kept for further consideration. This operation is achieved by using a look-up table.

The outcome of these four stages is a new image containing objects representing defects in the original image. Figure 6.10(e) displays the objects found by applying the new method to the acquired image (image A). The evidence of each object being a defect is computed by summing its grey levels. Then, any object with an evidence factor that is below a threshold (empirically set at 100) is removed.

Experiments were carried out with 75 images of birch wood boards, acquired with a CCD camera and both front and back lighting. The images were of size 512x512 pixels and had 256 grey levels. The method was implemented on an industrial vision system. 93% of defects were located correctly. The major advantage of the method is its speed because it employs functions based on dedicated parallel hardware rather than pixel-by-pixel software processing.

The total time for the whole process was 0.71 seconds on the Performance Vision Pi030 image-processing hardware used [Performance Vision, 1992]. Table 6.3 shows the execution time for each stage of the new process. The most time-consuming part was stage 2. This involved a 7x7 convolution and took 48% of the total processing time. To improve the speed further, a 5x5 or 3x3 convolution could replace the 7x7 convolution. Due to the rapid advances in image processing hardware, newer image processing systems would execute the method many times faster.

STAGE	TIME (seconds)
1	0.15
2	0.34
3	0.16
4	0.06
TOTAL TIME	0.71

Table 6.3 Execution time for each stage

After segmentation has been performed, two problems can be experienced. First, many areas, which are clear wood, may be detected as defects and, second, a defect may be represented by more than one segmented object (Figure 6.10(e)). This creates the need for an object processing stage after objects have been found. Pham and Alcock [1998c] developed two object-processing techniques to overcome these problems. The techniques are inspired by the artificial intelligence techniques of fuzzy logic and self-organising neural networks.

An algorithm that detects potential defects on wood may generate many objects of which not all will actually be defects. It is advantageous to eliminate these false objects at an early stage so that fewer clear wood objects are presented to the classification module. Therefore, a method was developed called *Accumulation of Evidence* that was inspired by fuzzy logic.

One of the main concepts in fuzzy logic is that statements do not have to be completely true (truth value = 1) or false (truth value = 0) but can have a degree of truth between 0 and 1. For example, the statement "This object is small" is neither true nor false but can be given an *evidence* value indicating the extent to which it is true. Most wood image segmentation techniques simply mark pixels as clear wood or defective. However, a pixel in an image of a birch wood sheet could be given an evidence value according to the expectation of that pixel representing a defect. Then, a pixel with a large evidence value is more likely to represent a defect than one with a small evidence value. Evidence values were generated by the segmentation method described above. In Figure 6.10, image D shows the values generated from image A. Darker pixels represent larger evidences.

After evidence values have been generated for all pixels, the detected pixels are grouped into objects. Pixels are placed into the same object if they are adjacent to one another in either the horizontal, vertical or diagonal directions. The *total evidence* for the object is calculated by summing the evidences of all the pixels in the object. If the total evidence of an object is below a specified threshold, then the object is removed from the segmented image. In experiments, the threshold employed was 100. In this way, small objects similar in shade to clear wood are removed. The technique will not remove an object if it is large or if its shade differs significantly from that of clear wood because its total evidence will be large.

After potential defects have been located and false objects removed, sometimes one defect will be represented by two or more objects (Figure 6.10(e)). If objects were to be presented to the classification module separately then it is likely that they would be incorrectly classified. One defect when split into smaller parts appears to be several smaller defects. The problem often happens with sound knots which contain some parts that are similar in grey-level to clear wood and so become segmented into several separate regions. A line object, such as a split, can appear as several disjointed objects. Also, a large defect may be segmented into one large object with several small objects.

A new method for clustering objects was proposed, inspired by the neural network architecture ART. In ART, the first pattern presented to the network creates a new neuron containing the attributes of the pattern. Subsequently, patterns are presented to the network and compared with those stored in existing neurons. If the new pattern is found to be *close* to a stored pattern then the neuron containing that pattern is updated so that its contents represent the new pattern and also all the patterns that were previously covered by the neuron. If the pattern is not deemed to be close to any of the stored patterns then a new neuron is created containing the attributes of that pattern. Whether a new pattern is close to a stored pattern depends upon a *distance function*.

The new object joining method has been implemented to operate in a similar manner to ART. First, four attributes are generated from the MBR co-ordinates of each object. These are derived from the lines that define the position of the MBR: start_x, end_x, start_y and end_y. Then, from these co-ordinates, the object's *area of interest* is found. This depends upon the MBR co-ordinates and the size of the MBR in the x and y directions.

To derive these, first, the lengths of the MBR in the x and y directions are calculated:

$$\text{len\_x} = \text{end\_x} - \text{start\_x} \qquad (6.1)$$

$$\text{len\_y} = \text{end\_y} - \text{start\_y} \qquad (6.2)$$

Next, the co-ordinates of the corners of the area of interest are determined. The co-ordinates are called aoi_sx, aoi_sy, aoi_ex and aoi_ey:

$$\text{aoi\_sx} = \text{start\_x} - (\text{len\_x} / 3) \qquad (6.3)$$

$$\text{aoi\_sy} = \text{start\_y} - (\text{len\_y} / 3) \qquad (6.4)$$

$$\text{aoi\_ey} = \text{end\_y} + (\text{len\_y} / 3) \qquad (6.5)$$

$$\text{aoi\_ex} = \text{end\_x} + (\text{len\_x} / 3) \qquad (6.6)$$

The value of 3, in the equations, was chosen empirically. The above equations have the effect of giving the area of interest the same proportions as the MBR. This means that an object with a long thin horizontal MBR will have a long thin horizontal area of interest. This is a useful feature because, for example, the area of interest of a line defect will also be a line but longer and wider than the MBR.

Figure 6.11 shows an object, its minimum bounding rectangle and its area of interest.

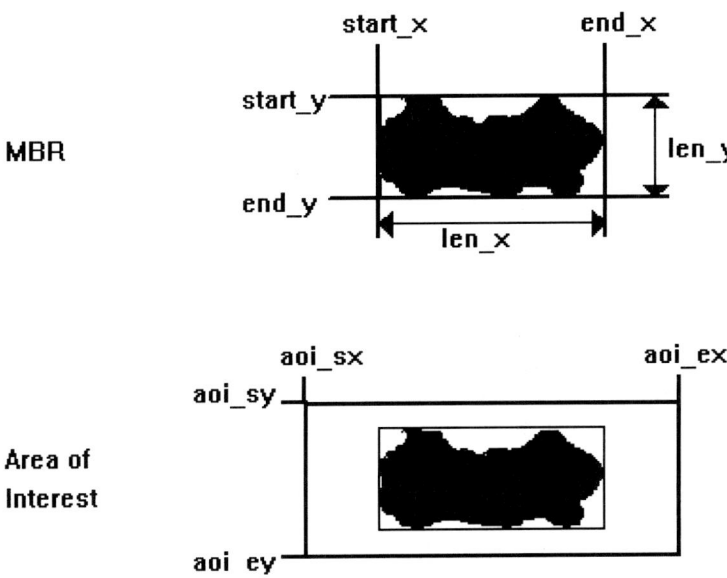

Figure 6.11 Minimum bounding rectangle and area of interest

Objects are presented to the clustering neural network in order of size (largest first) and tested to determine whether their MBR co-ordinates are close to the attributes of any existing neuron. Each neuron has four attributes (n_start_x, n_end_x, n_start_y and n_end_y) which define the position of its area of interest. The adopted distance function is such that if at least two of the corners of an object's MBR are within the area of interest of a neuron then the object is deemed close to the neuron and is assigned to it (Figure 6.12a).

The attributes of the neuron are then updated so that its area of interest includes the area that it previously covered and also the area of interest of the new object (Figure 6.12b). Then, the new area of interest of the neuron will be the smallest rectangle that can enclose the areas of interest of both the neuron as it was and the new object. If the new pattern is not close to any existing neuron (Figure 6.12c) then a new neuron is created with attributes equal to the co-ordinates of the four corners of the object's area of interest.

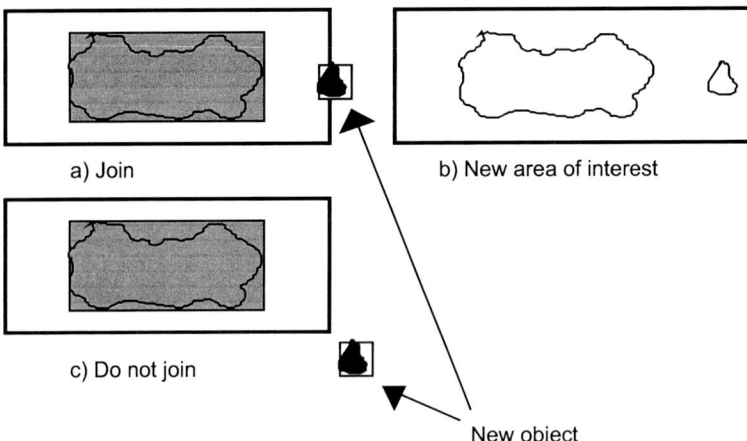

a) Join   b) New area of interest

c) Do not join

New object

Figure 6.12 Joining of objects

The object clustering technique is summarised below:

1. Sort objects into size order: largest first;
2. Assign the largest object to the first neuron, setting the attributes of the neuron equal to the co-ordinates of the corners of the object's area of interest;
3. Take the next largest object:

       repeat for each existing neuron:
       {
                   determine if the new object is *close* to the neuron
                   if it is *close*
                             then update the neuron and exit loop
       }
       if the object is not *close* to any existing neuron
                   then create a new neuron;
4. Go to (3) while there are still objects to cluster.

Tests were carried out on 75 images of birch wood boards. It was found that the average number of objects detected was reduced from 180 to 10 after the accumulation of evidence technique had been applied. No objects were removed that would negatively affect the overall grading result. The neural network joining method further reduced the average number of detected objects to seven. It was noted that the technique never joined together two defects of different types into the same region.

Figure 6.10(f) shows the result of applying the two techniques to the image in Figure 6.10(e). It can be seen that the problems of over and under segmentation have been overcome.

6.2.3 Subsequent Work

Other researchers have carried out follow-on work from this project. Drake and Packianather [1998] studied classification of wood defects using SCSs. Packianather and Drake [2000] developed methods for wood defect feature selection. Pham and Sagiroglu [2000] compared the MLP with LVQ for the classification of wood defects and found that the LVQ classifier gave a superior performance.

After the completion of the project, collaboration has taken place with North and South America. A partnership was formed with the University of Chile to develop a wood inspection system for the Chilean wood industry [Estevez et al, 1999]. The authors have also given advice to Ventek in Oregon, USA, in their development of a commercial veneer inspection system [Ventek, 2002]. Figure 6.13 shows the Ventek inspection system in a factory environment. A side view of the system with a view screen and conveyor system is shown in Figure 6.14. A meter is at the rear of the scanner to provide moisture information.

More recently, the authors have collaborated with the University of Patras in Greece [Stojanovic et al., 2001]. In this work, a six camera system was set up. Four cameras are employed to scan each of the four sides of the board. The other two cameras analyse the board at an inclined angle with the aid of laser line lighting. Two segmentation methods have been developed. The first, for biological defects such as knots, uses adaptive thresholding, smoothing and morphology. The second, for mechanical defects such as cracks, employs thresholding and profiling. Five features are extracted from the segmented objects: object height, object width, height/width ratio, position and compactness. Classification is performed by a fuzzy rule-based system.

Figure 6.13 Ventek GS2000 inspection system (Image courtesy of Ventek, Inc.)

Figure 6.14 Side view of GS2000 (Image courtesy of Ventek, Inc.)

6.2.4 Colour and Computer Tomography Inspection of Wood

To improve segmentation accuracy, it is possible to use colour vision instead of grey-scale images. Other researchers have performed experiments on using colour images for automated inspection of wood boards. Conners et al. [1985] utilised four channels (grey-scale, red, green and blue) to discover which combination would give the best results. It was found that the best performance was obtained by utilising all four channels together (76%). Red, green or blue individually gave a performance of around 58%. Grey-scale alone also produced 58% accuracy. The best two-colour combination was red and blue, which gave a result of 70%. Red, green and blue combined yielded 71%. Employing all four channels would impose a heavy computational burden on the system and so the use of grey-scale, blue and red was suggested.

Brunner et al. [1990] explained that one of the major difficulties in using colour vision is the increased computational overhead. Additionally, colour cameras are produced specifically for human viewers and so are not ideally suited to image processing. However, Funck et al. [1990] found that colour vision could detect streaks with an accuracy of 93% whereas grey-level images could only achieve 70%. Marszalec and Pietikainen [1993] also concluded empirically that colour vision adds much important information and that it increases the efficiency and speed of the inspection process.

Kline et al. [1998] carried out experiments analysing the performance of a colour vision system on red oak boards. The performance of the system was relatively low because of the sensitivity of the image processing algorithms. Due to the variability of defect-free wood on red oak boards, a large number of defects were found that were not actually present. Thus, it was suggested that other sensing techniques should be adopted.

Efforts are continuing to develop computer tomography (CT) inspection devices for wood boards and logs. CT scanners are useful for inspecting wood as defects normally have a different density from non-defective areas of the board. Additionally, they can detect internal defects. However, Schmoldt et al. [2000] highlighted some of the obstacles to the widespread adoption of CT scanning in the breakdown of hardwood logs into lumber. First, the amount of material that must be inspected is prohibitively high. Second, wood has a relatively low value compared to other objects that have employed AVI. Third, many wood processing factories are small and so do not have sufficient funds to purchase expensive equipment. However, CT scanning for wood is likely to become increasingly common in future.

6.3 Texture Classification

For textiles, a key property is their texture. Along with colour, textural information is useful in image segmentation. One of the most common applications of texture classification in AVI is in finding defects. A defective region appears as an area of the image where the texture is different from the rest of the image. One application is of texture classification in the inspection of leather for defects [Branca et al., 1996].

6.3.1 Classification using First-order Features

Cetiner [1995] carried out experiments on texture classification, initially employing only first-order features. The features that were utilised were mean, standard deviation, skewness, kurtosis, median, mode and inter-quartile range. The three classifiers used were the MLP, LVQ and rule sets derived by the RULES-3 algorithm [Pham and Aksoy, 1995]. The LVQ version adopted was LVQ2 with a conscience mechanism, as proposed by Pham and Oztemel [1996]. In classification, LVQ obtained 89.3%, RULES-3 gave 86.7% and MLP achieved 84.4%. Next, SCSs were tested to try to improve upon the accuracy achieved with LVQ. The combination strategy of Pham and Oztemel [1996] was adopted. It was found that combining three MLPs increased the accuracy to 91.6%. However, integrating three LVQ networks increased the performance only marginally to 89.7%. Utilising MLP, LVQ and RULES-3 gave a performance of 93.0%. The highest accuracy of 93.6% was obtained by using two MLP networks and the RULES-3 classifier. Figure 6.15 displays the best SCS found.

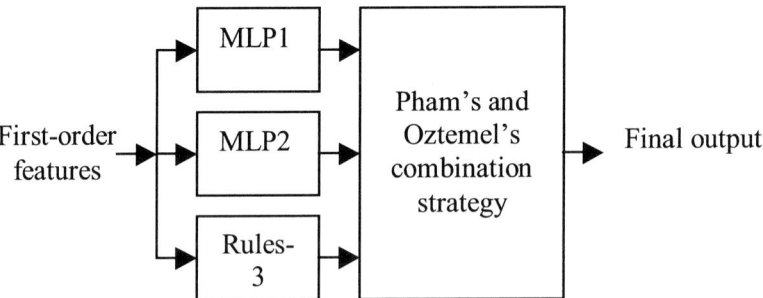

Figure 6.15 Synergistic classification system for texture recognition

Typically, texture recognition methods employ both first-order and second-order statistical features. It is possible for different textures to have the same first-order feature values if they have the same overall shade. Thus, second-order features are also required. It has been found that performance can be improved by adding first-order features to second-order information [Conners et al., 1983]. The same conclusions were reached by Balakrishnan et al. [1998] in the inspection of fabrics for defects.

6.3.2 Classification using Second-order Features

The most popular method of measuring second-order features in texture images is based on calculating Spatial Grey-Level Dependence (SGLD) or co-occurrence matrices [Haralick et al., 1973], as described in Chapter 4. The major problem with SGLD matrices is that they are computationally intensive.

Pham and Cetiner [1996] developed a new technique called Grey Level Difference (GLD) to calculate second-order features in a more reasonable time for real-time inspection. The result of the GLD technique is matrices, each element of which is the sum scaled difference in grey level between neighbouring pixels. To reduce the dimensions of the matrices, the grey levels are first quantised. As with SGLD, matrices can be calculated for each orientation and distance.

The GLD algorithm involves the following stages:

1. Assign the grey levels in the image into n quantisation levels. This reduces the size of the generated matrices and the associated computational burden. In an image originally containing Z grey levels, each quantisation level will cover roughly Z/n grey levels. For example, if the original image has 64 grey levels and the number of quantisation levels is set to five, the groups would be as follows: 0-12, 13-25, 26-38, 39-51 and 52-63. Each matrix would be of size 5x5.

2. Initialise all items in the nxn matrix to zero.

3. Select the first pixel in the image and label this as pixel 1. This pixel has a grey level p_1.

4. Find the neighbour of pixel 1 with the specified inter-pixel distance and direction. This is called pixel 2 and has a grey level p_2.

5. Scale the values of p_1 and p_2 to be between 0 and 1.

$$P_1 = \frac{p_1}{n}$$

$$(6.7)$$

$$P_2 = \frac{p_2}{n}$$

$$(6.8)$$

6. Calculate the difference between the grey levels of pixels 1 and 2 and increment this:

$$GLD = 1 + |P_1 - P_2|$$
(6.9)

The increment is performed to differentiate pixels with the same grey level from zero entries in the matrix. The value of GLD will be a number between 1 and 2. Its value will be closer to one when the two grey levels being compared are similar.

7. Round P_1 and P_2 to their nearest integer value. These two values correspond to a position in the matrix. P_1 represents the row number and P_2 the column.

8. Update the appropriate element in the GLD matrix:

$$new\_GLD(P_1, P_2) = old\_GLD(P_1, P_2) + GLD$$
(6.10)

9. If all the required distances and directions have been processed for the given pixel then go to step 10. Otherwise return to step 4.

10. If all pixels have been processed then terminate, otherwise return to step 3.

The method will now be explained with the aid of an example. Figure 6.16(a) shows a 4x4 window with pixel grey level values, where n equals 64. The number of quantisation levels is five.

Let Pixel 1 be element (3,3) with grey level 48 and Pixel 2 be element (3,4) with grey level 35. The scaled grey levels for these two pixels are $P_1=0.75$ (48/64) and $P_2=0.547$ (35/64). The GLD value is then $1 + |0.75-0.547| = 1.203$. p_1 falls in quantisation level 4 and p_2 lies in level 3, so the corresponding GLD matrix element is (4,3). Therefore, the value 1.203 is added to position (4,3) in the GLD matrix. Similarly, for elements (4,1) and (4,2) in the original image, the resulting GLD value is 1.25. Again, element (4,3) is to be updated in the GLD matrix. The resulting value at point (4,3) is 2.453, as can be seen in Figure 6.16(b). The rest of the GLD values are calculated similarly and the resulting matrix can be seen in Figure 6.16(b).

63	63	58	55
60	57	50	50
58	55	48	35
48	32	30	30

(a)

Interval	1	2	3	4	5
1	0	0	0	0	0
2	0	0	0	0	0
3	0	0	2.031	0	0
4	0	0	2.453	1	0
5	0	0	0	2.219	5.219

(b)

Intervals: 1 - Very dark pixels (grey levels 0 - 12); 2 - Dark pixels (grey levels 13 - 25); 3 - Medium pixels (grey levels 26 - 38); 4 - Light pixels (grey levels 39 - 51); 5 - Very light pixels (grey levels 52 - 63).

Figure 6.16 (a) Image window (b) GLD matrix calculated from the image [Cetiner, 1995]

The top left corner of the GLD matrix represents dark pixels and the bottom right corner covers bright pixels. Thus, for a dark image, larger values would be expected in the top left corner.

To test the technique, GLD matrices were calculated for sixteen texture images from the standard Broadatz album [Broadatz, 1968]. Broadatz textures can now be found on the Internet [University of Michigan, 2002]. The images were of size 128x128 and contained images of natural objects or scenes. Examples of the textures are shown in Figure 6.17.

Each image was divided into 32x32 non-overlapping rectangles, giving a total of 256 patterns. Images originally consisted of 256 grey levels but were quantised to eight levels. GLD matrices were calculated for each of the eight directions (every 45°) and for distances of d from one to five. Additionally, direction invariant matrices were calculated for each of the five distances by summing the matrices for that distance in all eight directions. Therefore, nine matrices were calculated for each of the five distances, giving forty-five matrices in total. The matrices were used directly as inputs for training the neural networks. Randomly, half were chosen for training and half for testing.

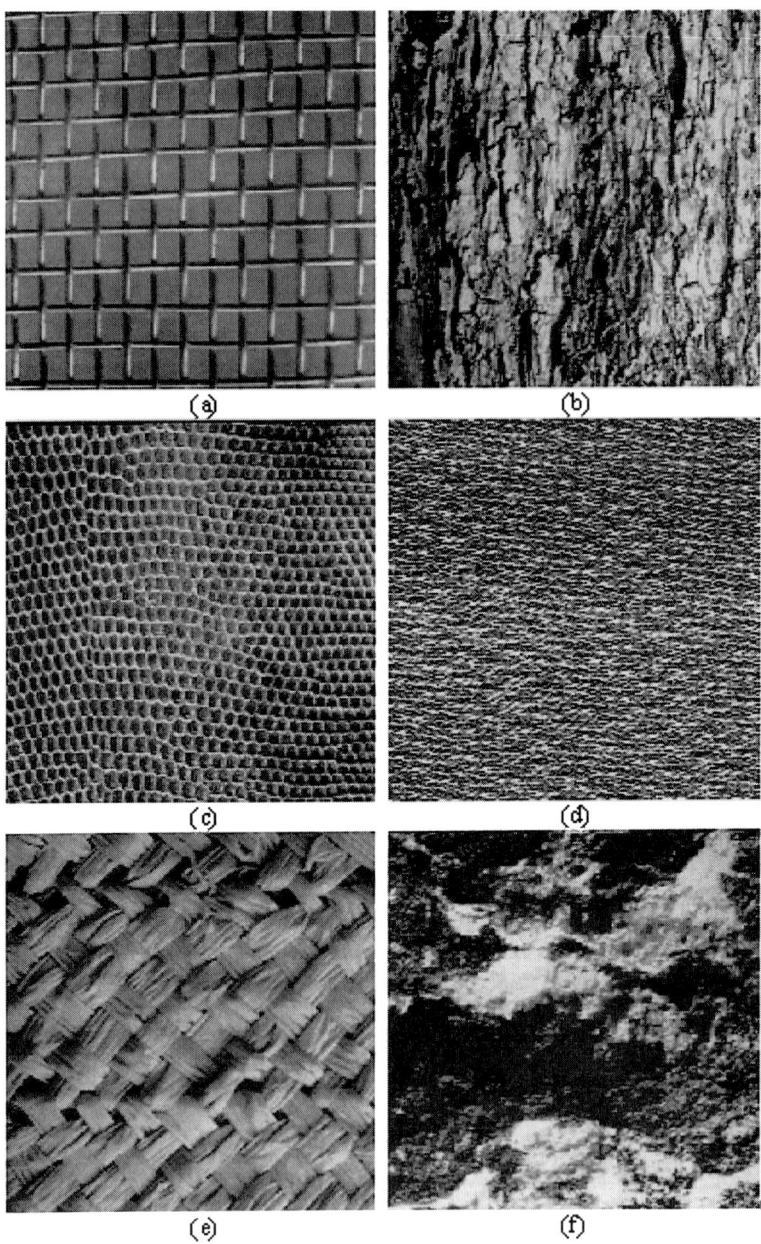

Figure 6.17 Example textures from the Broadatz album
(a) wire mesh (b) wood bark (c) reptile skin (d) cloth (e) basket weave (f) clouds

Two neural networks were selected for comparison: the LVQ2 network with a conscience mechanism and the MLP. In the experiments, the neural networks were implemented using the NeuralWorks Professional II software [NeuralWare, 2002]. The configurations of the LVQ and MLP networks are given in Tables 6.4 and 6.5 for GLD and SGLD, respectively. For SGLD, five features were calculated: energy, entropy, correlation, local homogeneity and inertia.

Number of neurons	MLP	LVQ2
Input	64 (8x8)	64 (8x8)
Hidden	39	96
Output	16	16

Table 6.4 Number of neurons for GLD (new technique)

Number of neurons	MLP	LVQ2
Input	5 (features)	5 (features)
Hidden	13	32
Output	16	16

Table 6.5 Number of neurons for SGLD (co-occurrence matrices)

Table 6.6 shows the results obtained. First, experiments were performed with a distance of one. Results showed that the GLD technique was better than the SGLD method for both MLP and LVQ2 classifiers. Also, it was found that the LVQ2 network performed significantly better than the MLP network. Next, the data sets were tested with distances two to five using the superior LVQ2 network. The GLD method was found to be significantly better than the SGLD technique for these distances.

		MLP	LVQ2
Distance 1	**SGLD**	67.2%	84.3%
	GLD	93.7%	96.0%
Distances 2 - 5	**SGLD**	-	80.3%
	GLD	-	94.6%

Table 6.6 Performance of SGLD and GLD techniques

6.3.3 Other Recent Work on Texture Classification

Co-occurrence matrices remain important in texture classification. For colour texture classification, Hauta-Kasari et al. [1999] used spectral co-occurrence matrices. Park and Chen [1996] adopted co-occurrence matrices for the inspection of poultry carcasses. By using a neural network, normal and abnormal carcasses could be separated without error.

Recent research into textural image classification has employed wavelets and lacunarity. For on-loom fabric inspection, Sari-Sarraf and Goddard [1999] developed a defect location algorithm based on the wavelet transform. The system gave a satisfactory classification rate under realistic operating conditions.

Lacunarity is derived from the study of fractals and is a measure of the gaps in the texture [Dong, 2000]. To calculate lacunarity, the image must be binarised, which can be performed using thresholding. Then, a sub-window of size RxR pixels is moved sequentially over the total image. From this, a histogram P(S) is generated, where P(S) represents the proportion of sub-windows which contain S white pixels. The mean M and variance V are calculated for P(S) and the lacunarity for value R is determined by:

$$L(R) = V/M^2 \qquad (6.11)$$

6.4 Summary

This chapter has described some of the applications of smart machine vision. These included inspection of car engine seals, wood boards and textured surfaces. Artificial intelligence techniques have been utilised successfully in all three applications.

References

Armstrong W.W. and Gecsei J. (1979) Adaptive Algorithms for Binary Tree Networks. *IEEE Trans. on Systems, Man and Cybernetics*. Vol. 9, pp. 276 - 285.

Balakrishnan H., Venkataraman S. and Jayaraman S. (1998) FDICS: A Vision-based System for the Identification and Classification of Fabric Defects. *Journal of the Textile Institute*. Vol. 89, No. 2, Part 1, pp. 365 - 380.

Branca A., Tafuri M., Attolico G. and Distante A. (1996) Automated System for Detection and Classification of Leather Defects. *Optical Engineering*. Vol. 35, No. 12, pp. 3485 - 3494.

Broadatz P. (1968) *Textures: A Photographic Album for Artists and Designers*. Van Nostrand Reinhold, New York.

Brunner C.C., Shaw G.B., Butler D.A. and Funck J.W. (1990) Using Colour in Machine Vision Systems for Wood Processing. *Wood and Fibre Science*. Vol. 22, No. 4, pp. 413 - 428.

Cetiner B.G. (1995) *Techniques for Texture Analysis and Classification*. PhD thesis, School of Engineering, Cardiff University, UK.

Conners R.W., McMillin C.W., Lin K. and Vasquez-Espinosa R.E. (1983). Identifying and Locating Surface Defects in Wood: Part of an Automated Lumber Processing System. *IEEE Trans. on Pattern Analysis and Machine Intelligence*. Vol. 5, No. 6, pp. 573 - 583.

Conners R.W., McMillin C.W. and Ng C.N. (1985) The Utility of Colour Information in the Location and Identification of Defects in Surfaced Hardwood Lumber. *Proc. 1ˢᵗ Int. Conf. on Scanning Technology in Sawmilling*. San Francisco, CA. pp. XVIII 1 - 33.

Dong P. (2000) Test of a New Lacunarity Method for Image Texture Analysis. *Int. Journal of Remote Sensing*. Vol. 21, pp. 3369 - 3373.

Drake P.R. and Packianather M.S. (1998) A Decision Tree of Neural Networks for Classifying Images of Wood Veneer. *Int. Journal of Advanced Manufacturing Technology*. Vol. 14, No. 4, pp. 280 - 285.

Estevez P.A., Fernandez M., Alcock R.J. and Packianather M.S. (1999) Selection of Features for the Classification of Wood Boards. *Proc. 9ᵗʰ Int. Conf. Artificial Neural Networks*. Edinburgh, UK. pp. 347 - 352.

Freeman H. (1974) Computer Processing of Line Drawing Images. *Computing Surveys*. Vol. 6, pp. 57 - 97.

Funck J.W., Brunner C.C. and Butler D.A. (1990) Softwood Veneer Defect Detection Using Machine Vision. *Proc. Symp. on Process Control / Production*

Management of Wood Products: Technology for the 90's.   Forest Products Research Society, Madison, WI. pp. 113 - 120.

Haralick R.M., Shanmugam K. and Dinstein I. (1973) Textural Features for Image Classification. *IEEE Trans. on Systems, Man and Cybernetics.* Vol. 3, No. 6. pp. 610 - 621.

Hauta-Kasari M., Parkkinen J., Jaaskelainen T. and Lenz R. (1999) Multi-spectral Texture Segmentation Based on the Spectral Co-occurrence Matrix. *Pattern Analysis and Applications.* Vol. 2, No. 4, pp. 275 - 284.

Huber H.A., McMillin C.W. and McKinney J.P. (1985) Lumber Defect Detection Abilities of Furniture Rough Mill Employees. *Forest Products Journal.* Vol. 35, No. 11/12, pp. 79 - 82.

Kline D.E., Widoyoko A., Wiedenbeck J.K. and Araman P.A. (1998) Performance of Color Camera Machine Vision in Automated Furniture Rough Mill Systems. *Forest Products Journal.* Vol. 48, No. 3, pp. 38 - 45.

Marszalec E. and Pietikainen M. (1993) Colour Analysis for Automated Visual Inspection of Pine Wood. *SPIE Vol. 1907: Machine Vision Applications in Industrial Inspection.* San Jose, CA. pp. 80 - 94.

NeuralWare (2002) *NeuralWare Professional II.* NeuralWare, Carnegie, PA. http://www.neuralware.com

Packianather M.S. and Drake P.R. (2000) Neural Networks for Classifying Images of Wood Veneer. Part 2. *Int. Journal of Advanced Manufacturing Technology.* Vol. 16, No. 6, pp. 424 - 433.

Park B. and Chen Y.R. (1996) Multispectral Image Co-occurrence Matrix Analysis for Poultry Carcasses Inspection. *Trans. of the ASAE.* Vol. 39, No. 4, pp. 1485 - 1491.

Performance Vision (1992) *Image Analysis Library Manual.* Performance Vision Ltd., Solihull, West Midlands, UK.

Pham D.T. and Aksoy M.S. (1995) A New Algorithm for Inductive Learning. *Journal of Systems Engineering.* Vol. 5, pp. 115 - 122.

Pham D.T. and Alcock R.J. (1996) Automatic Detection of Defects on Birch Wood Boards. *Proc. IMechE. Part E - Journal of Process Mechanical Engineering.* Vol. 210, pp. 45 - 52.

Pham D.T. and Alcock R.J. (1998a) Automated Grading and Defect Detection: A Review. *Forest Products Journal*. Vol. 48, No. 4, pp. 34 - 42.

Pham D.T. and Alcock R.J. (1998b) Recent Developments in Automated Visual Inspection of Wood Boards. In *Advances in Manufacturing Systems - Decision, Control and Information Technology*. (ed. Tzafestas S.G.). Springer-Verlag, Berlin and London. pp. 79 - 87.

Pham D.T. and Alcock R.J. (1998c) Artificial Intelligence Techniques for Processing Segmented Images of Wood Boards. *Proc. IMechE. Part E - Journal of Process Mechanical Engineering*. Vol. 212, pp. 119 - 129.

Pham D.T. and Alcock R.J. (1999a) Plywood Image Segmentation Using Hardware-Based Image Processing Functions. *Proc. IMechE. Part B - Journal of Engineering Manufacture*. Vol. 213, No. 4, pp. 431 - 434.

Pham D.T. and Alcock R.J. (1999b) Automated Visual Inspection of Wood Boards: Selection of Features for Defect Classification by a Neural Network. *Proc. IMechE. Part E - Journal of Process Mechanical Engineering*. Vol. 213, pp. 231 - 245.

Pham D.T. and Alcock R.J. (1999c) Synergistic Classification Systems for Wood Defect Identification. *Proc. IMechE. Part E - Journal of Process Mechanical Engineering*. Vol. 213, No. 2, pp. 127 - 133.

Pham D.T. and Bayro-Corrochano E.J. (1994) Neural Classifiers for Automated Visual Inspection. *Proc. IMechE. Part D - Journal of Automobile Engineering*. Vol. 208, pp. 83 - 89.

Pham D.T. and Bayro-Corrochano E.J. (1995) Neural Networks for Classifying Surface Defects on Automotive Valve Stem Seals. *Int. Journal Machine Tools Manufacture*. Vol. 35, No. 8, pp. 1115 - 1124.

Pham D.T. and Cetiner B.G. (1996) A New Method for Describing Texture. *Proc. 3rd Int. Workshop on Image and Signal Processing*. Manchester, UK. pp. 187 - 190.

Pham D.T. and Oztemel E. (1996) *Intelligent Quality Systems*. Springer Verlag, Berlin and London.

Pham D.T. and Peat B.J. (1995) Hybrid Method for Systems Analysis. *IEE Proc. Science and Measurement Technology*. Vol. 142, No. 5, pp. 350 - 361.

Pham D.T. and Sagiroglu S. (2000) Neural Network Classification of Defects in Veneer Boards. *Proc. IMechE. Part B - Journal of Engineering Manufacture.* Vol. 214, No. 3, pp. 255 - 258.

Pham D.T., Jennings N.R. and Ross I. (1995) Intelligent Visual Inspection of Valve-Stem Seals. *Control Engineering Practice.* Vol. 3, No. 9. pp. 1237 - 1245.

Polzleitner W. and Schwingshakl G. (1992) Real-Time Surface Grading of Profiled Wooden Boards. *Industrial Metrology.* Part 2, pp. 283 - 298.

Sari-Sarraf H. and Goddard J.S. (1999)   Vision System for On-loom Fabric Inspection. *IEEE Trans. on Industrial Applications.* Vol. 35, No. 6, pp. 1252 - 1259.

Schmoldt D.L., Occena L.G., Lynn Abbott A. and Gupta N.K. (2000) Nondestructive Evaluation of Hardwood Logs: CT Scanning, Machine Vision and Data Utilization Nondestructive Testing and Evaluation. *Nondestructive Testing and Evaluation.* Gordon and Breach, Newark NJ.

Stojanovic R., Papadopoulos G., Mitropoulos P., Alcock R.J. and Djurovic I. (2001) An Approach for Automated Inspection of Wood Boards. *Int. Conf. on Image Processing.* IEEE Signal Processing Society. Thessaloniki, Greece. pp. 798 - 801.

University of Michigan (2002) *Broadatz Textures.* University of Michigan, College of Engineering, Robert H. Lurie Engineering Center, Ann Arbor, MI. ftp://freebie.engin.umich.edu/pub/misc/textures

Ventek (2002) *GS2000 Veneer Inspection System.* Ventek Inc., Eugene, Oregon, USA. http://www.ventek-inc.com

Problems

1. Read the description of the IMPRINT project on the attached ProVision CD. Perform a short literature search to find at least one example of a similar application.

2. In pseudo-code, write an algorithm to perform the IMPRINT inspection task.

3. Read the following sections in the ProVision online manual:
- I Digital Output;
- I Image Capture;

- I Serial Output;
- L Intersection Point/Angle;
- L Position Generation;
- M Arithmetic;
- M Decision Making;
- V Pattern Correlation;
- W Binarization;
- W Blob Analysis;
- W Blob Selection;
- W Distance Profile;
- W Dominant Edge.

Write a short summary for each function, explaining what it does.

4. Using the functions described in Problem 3 (or others if you prefer), write an inspection program to perform the IMPRINT inspection.

5. Test the program that you have written on the images PRINT1.BMP, PRINT2.BMP and PRINT3.BMP given on the attached CD.

6. Open the IMPRINT project and run the program given. Compare the results of your algorithm with that of the sample program.

Chapter 7

Industrial Inspection Systems

This chapter describes typical functions of industrial vision systems, overviews some state-of-the-art vision units and discusses possible future trends in AVI.

7.1 Image Processing Functions of Inspection Systems

Many inspection systems are based on standard personal computers with a frame-grabber card installed. A camera is attached to the card using a cable and images are stored and manipulated on the computer. Advantages of this approach, compared with dedicated vision systems, are that personal computers are inexpensive and also familiar to many people.

Inspection systems based on a personal computer require special software to perform the image analysis. Two established software packages for image processing are Visilog and Khoros. Visilog5 is a 32-bit image processing package [Noesis, 2002]. It can run on the Unix and Windows NT operating systems and can operate together with a wide range of frame grabbers. Khoros Pro 2001 has a Windows®-based graphical user interface [Khoral, 2002], which offers features such as drag-and-drop icons and multiple working windows. A main feature of Khoros is its high-level visual programming language called Cantata. The Cantata language employs flow diagrams to describe data inputs, outputs and processing. Therefore, it is easy to use and intuitive. However, if preferred, Khoros commands can be executed from the command prompt.

The Siemens ProVision software, demonstrated on the attached CD, is a graphical Windows-based development environment. ProVision includes functions for image capture, program flow control and communication with interfaces. Functions are available to process windows, lines, vectors, points and circles. For example, the window functions cover:

- Addition, subtraction and logical combination of windows;
- Convolution filters;
- Edge detection;
- Feature extraction (blob analysis);
- Histogram calculation and smoothing;
- Morphological erosion and dilation;
- Pattern search;
- Profiling;
- Thresholding.

7.2 State-of-the-Art Vision Systems

An example of an industrial AVI system is the Siemens VS710 smart camera [Siemens, 2002]. The VS710 has a Profibus interface so that it can be connected to other cameras, PCs and programming devices. Images and programs can be transmitted to the camera from remote locations via the Profibus network. The system can operate at speeds of up to 25 pieces per second. Applications of the VS710, given on the attached CD, include:

- Bolt - position determination of threaded bolts;
- Code - reading of bar-type codes;
- Contact - measurement of distance between contacts;
- Date - checking presence and completeness of printed characters;
- Imprint - inspection of printed symbols on plastic parts;
- Label - label inspection on bottles;
- Pack - detection of correct cover type on packages;
- Sort - bottle sorting according to bottle shape;
- Washer - surface defect detection.

See Appendix and attached CD for more details of ProVision.

Features of modern inspection systems include advanced cameras, intuitive development environments, intelligent algorithms and high performance. These features will now be discussed in more detail.

7.2.1 Advanced Cameras

A recent development in AVI is the intelligent camera, alternatively known as the smart camera. The term "intelligent" is applied to cameras that contain a processor chip and so can perform image processing inside the camera instead of remotely. In other words, an intelligent camera is an integrated sensor and processing unit.

DVT Corporation (together with the Georgia Institute of Technology) pioneered the smart camera [Hardin, 2000]. Figure 7.1 shows the standard DVT Series 600 SmartImage Sensor [DVT, 2002]. The latest DVT smart camera is the Legend 530 SmartImage Sensor. The camera, shown in Figure 7.2, can be fitted easily into restricted spaces and comes with the option of an integrated light source. It can be configured with several different processors to meet the customer's requirements and budget.

Figure 7.1 DVT Series 600 SmartImage Sensor (image courtesy of DVT corporation)

There are now many manufacturers producing smart cameras. Zuech [2001] identified the advantages and disadvantages of smart camera technology compared with "embedded systems". The main advantage of smart cameras is that they are more economical and so can be applied to areas in which otherwise machine vision would be too expensive. Their major disadvantage is that they do not have the same processing power and functionality as embedded systems. However, smart cameras will become more powerful in future and it is expected that the performance gap will decrease.

Figure 7.2 DVT Legend 530 SmartImage Sensor (image courtesy of DVT Corporation)

The rise in importance of smart cameras has been highlighted by Hardin [2000], who states that many smart camera manufacturers are now offering 2D bar code reading and character recognition capabilities in their cameras. The next phase for intelligent cameras is predicted to be the use of multiple intelligent camera systems connected across a network. In this way, inspection of a complicated product can be broken down into less complex sub inspections. A secondary advantage of this is that faults will be detected earlier in the manufacturing process, before value has been added to already-defective products.

Due to their performance and relative low cost, smart cameras are quickly replacing frame-grabber inspection systems [Williams, 2001]. Many smart cameras will offer colour inspection, together with higher speeds and resolutions, which will increase their potential applications. In addition, smart cameras with Ethernet capabilities will allow them to be connected to local-area networks and the Internet. This will enable remote programming and monitoring of smart cameras.

Cameras are available that can capture images with wavelengths different from that of light. Normal cameras have a peak sensitivity of around 500 nanometers. Hitachi Denshi [2002] produces cameras with a spectral response extending to the near infrared region. The Hitachi KP-MR2 has a peak sensitivity of 640 nanometers and a useful sensitivity surpassing 900 nanometers. The Hitachi KP-F2 gives a peak sensitivity of 760 nanometers and more than 1000 nanometers for useful sensitivity.

Pulnix [2002] are developing ultra-violet sensitive progressive-scan CCD Cameras. Such devices eliminate the need for special blocking materials in front of the lens, such as coating, plastic and glass. Applications for the camera include surface and laser inspection.

For 3D applications, IVP produce a camera that can acquire 3D images directly [IVP, 2002]. The camera outputs line profiles that are combined to create a 2D image containing height information. The intensity of a pixel represents the height of the object at that point.

7.2.2 Intuitive Development Environments

Many industrial vision systems now adopt Windows-based programming methods, which most users find easier than systems employing text-based programming. Windows-based systems have features such as pull-down menus, buttons and text input boxes.   The visual approach to programming facilitates application development because it is well known to most computer users. An example of a Windows-based AVI system is the Pulnix ZiCAM [Pulnix, 2002]. Figure 7.3 shows a typical screen view, containing many of the familiar windows features.

The development environment of the Cognex In-Sight 2000™ vision sensor, shown in Figure 7.4, is spreadsheet based [Cognex, 2002]. Many users are now familiar with this form of data entry and so configuring vision applications is fast and simple. The PPV+ (Print Presence Verification Plus) packaging application from the Cognex In-Sight range is for print presence inspection and print verification on products. It features a "Train and Go" graphical user interface so that development can be performed without programming.

Braintech's eVisionFactory development environment is based on Microsoft's COM (Common Object Model) components [Braintech, 2002].   COM components have the advantages of being re-usable and of containing clear documented interfaces. Another feature of eVisionFactory is electronic online support. The first level of support is trouble-shooting decision trees that result in video-based help. If this is not sufficient, connection can be made to a support engineer who will advise the user with the aid of voice chat and white boarding. If the problem is still not solved, the remote engineer can connect to the system and perform trouble-shooting as if working on a local machine.

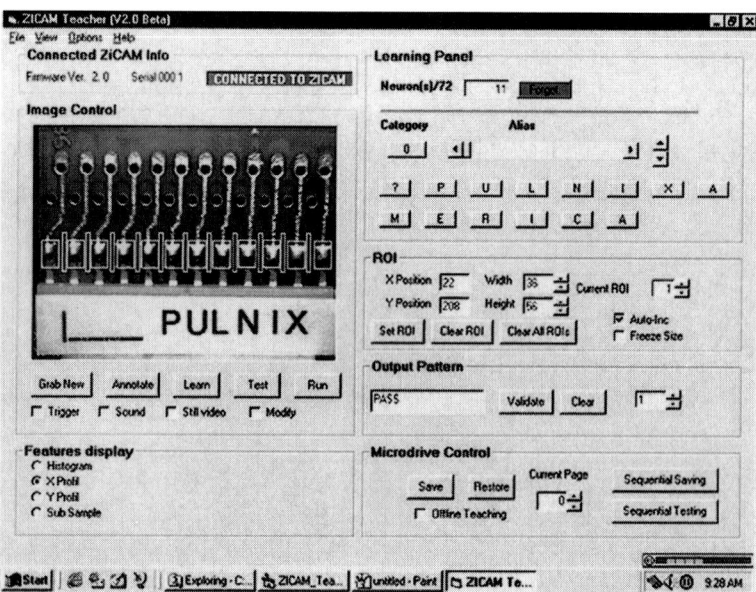

Figure 7.3 Windows-based user interface (image courtesy of Pulnix America, Inc.)

7.2.3 Intelligent Algorithms

Recently, artificial intelligence techniques have been incorporated into several industrial inspection systems.

The PULNiX ZiCAM™ [Pulnix, 2002] is an intelligent camera, which has a Windows-based user interface and contains hardware neural network technology. The ZiCAM is shown in Figure 7.5. ZiCAM is an acronym for Zero instruction CAMera, which means that it does not need programming but is simply presented good and bad parts and learns how to differentiate them using the neural network. ZiCAM captures an image of the product and passes it to a module called MUREN (MUltimedia Recognition ENgine). Several features are extracted by MUREN, including the histogram, profiles and pixel subsamples. The feature vectors extracted consist of 64 features. These vectors are passed to two processors, which contain the neural network hardware. The neural networks together have 74 outputs, so instead of giving just a "PASS/FAIL" output, the ZiCAM can be trained to separate products into up to 74 classes.

Smart Search is an adaptive pattern-locating tool, which is provided by Coreco Imaging as part of their Sherlock and MVTools vision packages [Coreco, 2002]. Smart Search can be used for inspection tasks such as assembly, print and character verification. It contains a Training Wizard that is based on artificial intelligence to facilitate system training. The developer need only provide good and bad examples of the object to be inspected and then Smart Search learns the characteristics of that object automatically. Figure 7.6 shows how robust Smart Search is to image degradation. It can cope with image blur, noise, rotation poor contrast and incomplete parts.

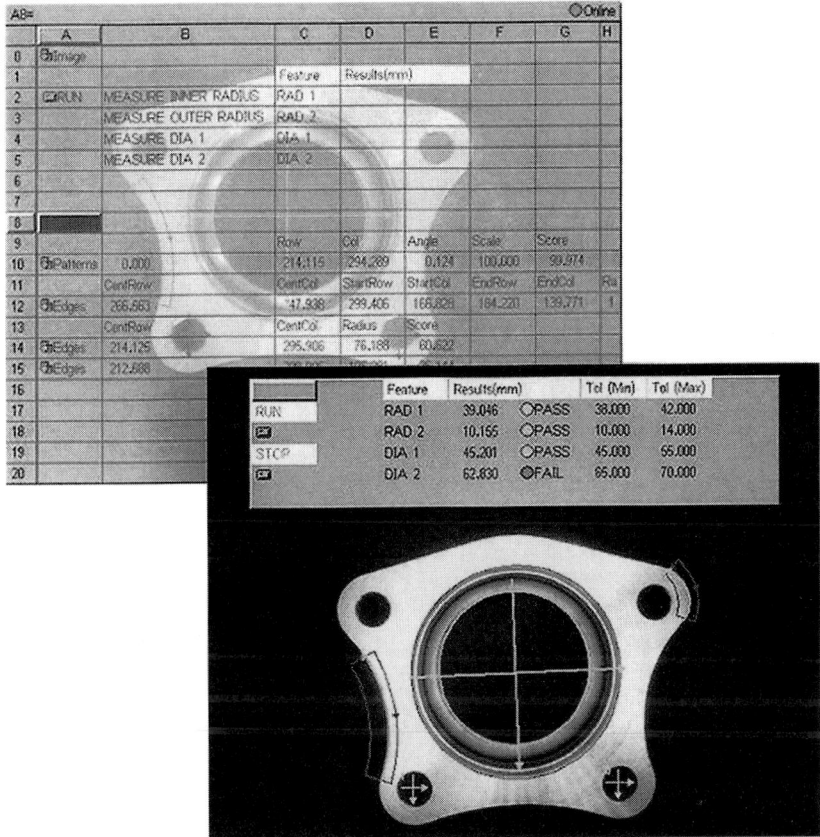

Figure 7.4 Spreadsheet development environment (image courtesy of Cognex Corporation)

Figure 7.5 ZiCAM camera (image courtesy of Pulnix America, Inc.)

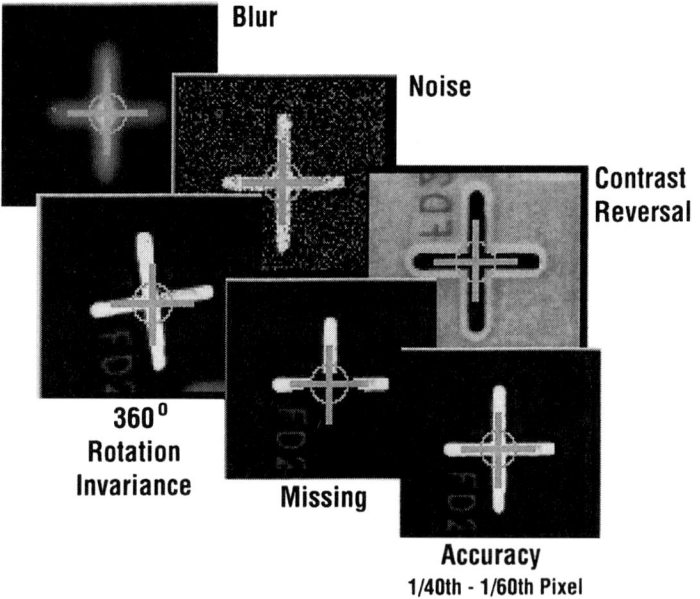

Figure 7.6 Robustness of Smart Search (image courtesy of Coreco Imaging)

The NeuroCheck Compact® is an intelligent camera that can be programmed using point and click techniques, rather than having to write and compile programs [DS, 2002]. The system integrates a camera, processor, I/O and LAN connection in a single compact housing and has the full functionality of the Windows-based Neurocheck package [Demant et al., 1999]. This package is able to compute a number of object features from regions of interest in the image. These features are then passed to the classification module, which determines the type of the region being inspected.

The Sightech Eyebot does not need programming but learns part form and features using a combination of fuzzy logic and neural networks, known as Neuro-Ram [Sightech, 2002]. The Eyebot is available in two formats, the Shape Eyebot and the Spectrum Eyebot. The Shape Eyebot recognises the shape of objects that are placed in front of the camera and can then detect any which deviate from the learned shape. The Spectrum Eyebot learns the colour of objects that are shown to it. The output of the system can be an integer from 0 to 99, indicating the status of the product, or it could be an optical output to be passed to a PLC. Suggested applications for the Spectrum Eyebot are inspecting fruit, determining whether potato chips are burnt, checking colour labels, ascertaining whether grass needs watering and verifying that no chickpeas are mixed in with corn.

Inventions [2002] incorporate many artificial intelligence techniques into their ILIB software. ILIB contains libraries with image analysis and artificial intelligence functions. ILIB1 is the image-processing library and ILIB2 is designed for image analysis. ILIB3 contains pattern recognition, statistical, fuzzy logic and neural network tools. The neural network functionality of ILIB3 is provided by a multi-layer neural network with many configurable parameters, including learning rate, momentum, number of neurons, number of layers and activation functions. ILIB3 also contains pattern recognition techniques, such as minimum distance and K-nearest neighbour classification. Additionally, the distributions of feature vectors can be analysed by many statistical techniques such as ANOVA. The software also incorporates fuzzy logic classification functions.

The intelligent techniques of ILIB3, shown in Figure 7.7, include genetic imaging [Hickey, 1997]. In this method, the user provides the system with the original image and the image that is desired from image processing. Then, ILIB3 uses genetic algorithms to derive the convolution filter or sequence of filters required to convert the original image to the target one.

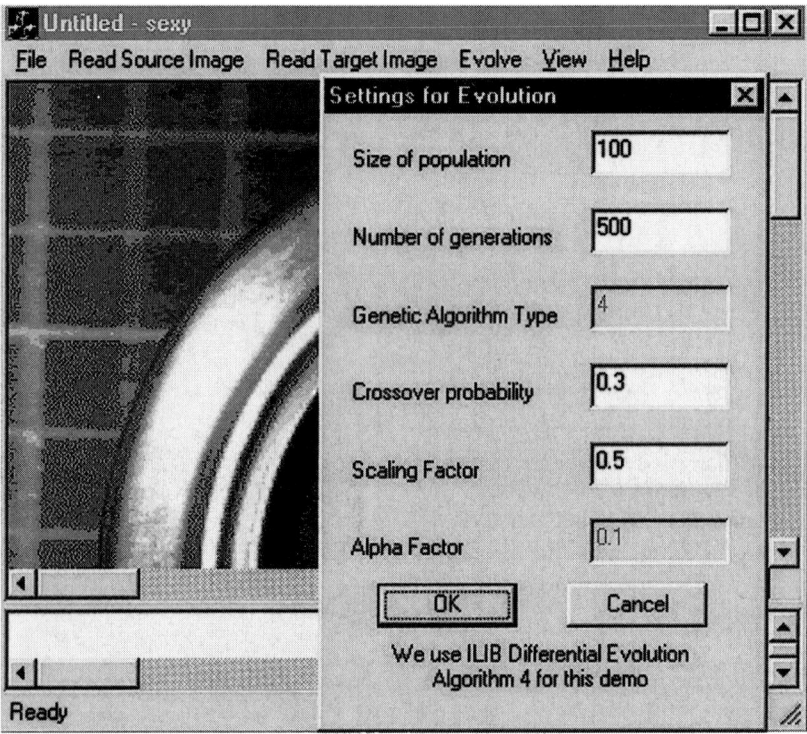

Figure 7.7 ILIB3 genetic imaging library (image courtesy of Inventions Ltd.)

The software of Braintech Inc. employs a wide range of artificial intelligence techniques, such as neural networks, fuzzy logic, natural language processing, qualitative maths and genetic algorithms [Braintech, 2002]. The software has been employed for inspection tasks such as mould number recognition, brake shoe identification and positioning of cylinder heads for the automotive industry (Figure 7.8). Discussing Braintech's software, Wright [2001] states that "Artificial Intelligence-based analysis in vision offers unprecedented flexibility and exploitability in industry".

7.2.4 High Performance

Industrial inspection systems often need to perform complex processing in a restricted time. For example, production lines where bottles are filled and labelled can operate at speeds up to 2000 bottles per minute. Continual industrial advances mean that camera frame rates and processor speeds are increasing rapidly.

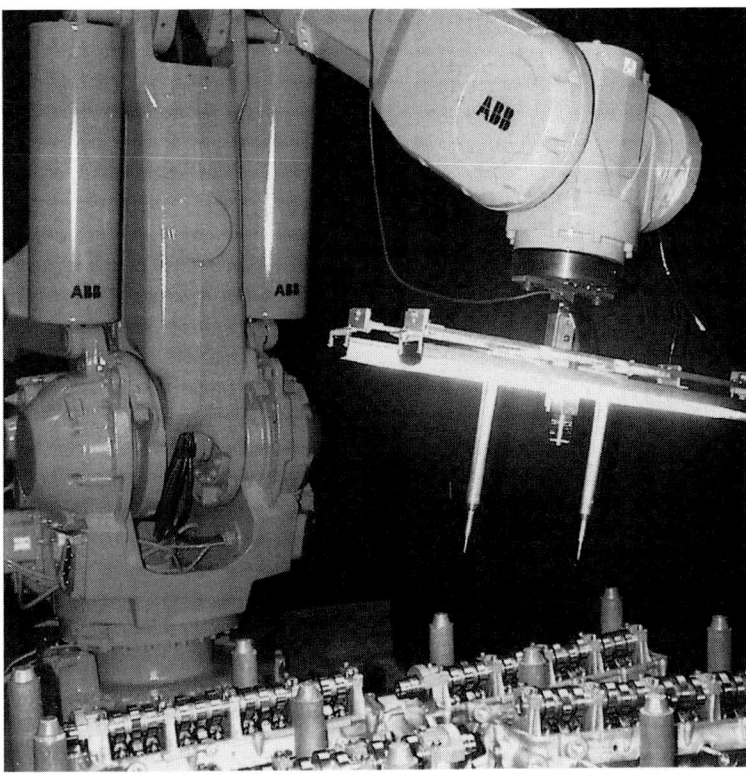

Figure 7.8 Braintech's vision-guided robotics (image courtesy of Braintech Inc.)

Two methods for increasing inspection speed are digital signal processing (DSP) boards and parallel processing systems. A number of companies now exist that can provide these technologies for inspection applications.

DSP hardware is able to perform image-processing operations in a fraction of the time that would be required by software. Kane Computing [2002] offers a wide range of DSP equipment for developing fast inspection systems, including starter kits, evaluation modules, emulators and boxed systems.

Azure [2002] offers a range of parallel processing vision systems. The parallel processing of the Azure AVS101 can operate in two ways. First, an algorithm can be run on several processors so that many images can be processed simultaneously. Second, each processor could run a different part of the algorithm so that one image can be analysed more quickly. The AVS101 offers scaleable processing, meaning that if the speed is not sufficient, more processors can be added. Figure 7.9 displays the Azure AVS101 system.

Figure 7.9 Parallel processing inspection system (image courtesy of Azure Ltd.)

7.3 Future of Industrial Inspection Systems

There is a growing acceptance of AVI systems in industry. The Automated Imaging Association (AIA), the trade group of the North American machine vision industry, stated that the total sales for machine vision equipment increased by 26% in the year 2000 [AIA, 2001]. The worldwide market for machine vision systems was calculated to be around $6.2 billion. This was divided into $2.1 billion for North America, $1.9 billion in Japan, $1.5 billion for Europe and $0.7 billion in the rest of the world. It is predicted that the global machine vision market will reach around

$12 billion by 2005. Rooks [1997; 1998; 1999; 2000; 2001] has also observed that interest in machine vision systems from the manufacturing sector has been gradually increasing in recent years.   The AIA estimates that only 10% of possible applications have been met.

Analysing current trends and problems in AVI can give some insight into its future. Wagner [2001] views the progress of industrial inspection systems in terms of four pillars:
1. faster;
2. cheaper;
3. more accurate;
4. more robust.
In addition, there is a continuing effort to create development environments that are user-friendly.

A current limitation of vision systems is that they work in restricted environments, where the lighting and object positions are well defined. This may be solved in the future but for industry, the important factor is that the vision systems operate accurately and not that they can function in all lighting conditions. Therefore, even though progress is expected in this area, the solution to this problem is not of critical importance to industry.

A second obstacle to the widespread adoption of AVI has been the cost of development, implementation and maintenance of the systems.   Fortunately, the performance-to-price ratio of machine vision systems is increasing rapidly and the range of applications to which they can be applied will expand.

Perhaps the greatest challenge for AVI systems is that today's manufacturing environment is characterised by products that change rapidly.   For the widespread acceptance of automated inspection, Wagner [2001] highlights two important points. First, algorithms need to be adaptive so that they can be easily tailored to new products.   Second, inspection systems should become more intuitive so that a specialised engineer is not needed to program and maintain them. For both of these objectives, Wagner proposes artificial intelligence as the solution. This trend has already started, as inspection systems are now available that employ AI to facilitate configuration as well as movement from one inspection task to another.

7.4 Summary

This chapter has reviewed the state-of-the-art in industrial inspection system technology. Recent developments include advanced cameras, intuitive development environments, intelligent algorithms and high performance.

An important area of progress in industrial inspection systems is artificial intelligence. The number of systems that employ AI to facilitate development and reconfiguration is increasing. Artificial intelligence is set to play a key role in the future of automated inspection.

References

AIA (2001) *The Machine Vision Market: 2000: Results and Forecast to 2005*. Automated Imaging Association. Ann Arbor, MI.

Azure (2002) *Azure Ltd*. Dorchester, UK. www.azure.com

Braintech (2002) *Braintech Inc*. North Vancouver, BC, Canada. www.braintech.com

Cognex (2002) *Cognex Corporation*. Natick, MA. www.cognex.com.

Coreco (2002) *Coreco Imaging Inc*. St. Laurent, Quebec, Canada. www.imaging.com

Demant C., Streicher-Abel B. and Waszkewitz P. (1999) *Industrial Image Processing: Visual Quality Control in Manufacturing*. Springer-Verlag, Berlin.

DS (2002) *Datenverarbeitung und Sensortechnik GmbH*. Remseck, Germany. www.neurocheck.com.

DVT (2002) *DVT Corporation*. Norcross, GA. www.dvtsensors.com

Hardin W. (2000) Smart Cameras: The Last Step in Machine Vision Evolution? *Machine Vision Online*. www.machinevisiononline.org

Hickey D.S. (1997) Genetic Imaging - The End of Systems Integration. *Image Processing Europe*. October, pp. 20 – 23.

Hitachi Denshi (2002) *Hitachi Denshi America Ltd.* Woodbury, NY. www.hdal.com.

Inventions (2002) *Inventions Ltd.* Sale, Cheshire, UK. www.inventions.u-net.com

IVP (2002) *IVP AB.* Linköping, Sweden. www.ivp.se

Kane Computing (2002)   *Kane Computing.* Northwich, Cheshire, UK. www.kanecomputing.com

Khoral (2002) *Khoros.* Khoral Research, Albuquerque, New Mexico, USA. www.khoral.com

Noesis (2002) *Visilog 5.* Noesis S.A., Cedex, France. www.noesisvision.com

Pulnix (2002) *PULNiX America Inc.* Sunnyvale, CA. www.pulnix.com

Rooks B.W. (1997) Vision Arrives at Manufacturing Week. *Sensor Review.* Vol. 17, No. 1, pp. 33 - 37.

Rooks B.W. (1998) Vision Applications Emerge at Manufacturing Week. *Sensor Review.* Vol. 18, No. 2, pp. 88 - 91.

Rooks B.W. (1999) Vision Again the Star at Manufacturing Week. *Industrial Robot.* Vol. 26, No. 2, pp. 115 - 120.

Rooks B.W (2000) A Good Week for Manufacturing. *Assembly Automation.* Vol. 20, No. 1, pp. 46 - 51.

Rooks B.W. (2001) Manufacturing Week Targets Assembly and Vision. *Assembly Automation.* Vol. 21, No. 2, pp. 117 - 122.

Siemens (2002) *Siemens UK.* Sir William Siemens House, Manchester, UK. www.ad.siemens.de/machine-vision

Sightech (2002) *Sightech.* San Jose, CA. www.sightech.com

Wagner G. (2001) Emerging Trends in Machine Vision. *Machine Vision Online.* www.machinevisiononline.org

Williams M.E. (2001) Smart Cameras Increase Their Intelligence - The Latest Trends in Vision Sensor Technology. *Machine Vision Online.* www.machinevisiononline.org

Wright D.T. (2001) A View to the Future: Braintech's Vision and Next Generation Manufacturing. *Machine Vision Online*. www.machinevisiononline.org

Zuech N. (2001) Are Smart Cameras Smart Enough? *Machine Vision Online*. www.machinevisiononline.org

Problems

1. Carry out an updated review of the state-of-the-art in commercial vision systems.

2. Find up-to-date figures on the current growth in the market for machine vision systems. How do the new figures compare with those given in this chapter?

3. The evolution of machine vision systems has gone from dedicated systems, through systems based on personal computers, to smart cameras. Find out if smart cameras are still the last step in machine vision evolution.

4. Search for new examples of machine vision systems that employ artificial intelligence. Identify the type of artificial intelligence technique employed in each system.

5. Take two state-of-the-art machine vision systems and compare them in terms of camera capabilities, user friendliness, image processing functionality and their respective performances.

6. Read the section on learning in the ProVision online manual. Describe how learning operates in ProVision.

Appendix

Siemens ProVision Software Package

Overview

ProVision is a Windows®-based development environment for the Siemens SIMATIC VS710 PROFIBUS DP Intelligent Vision Sensor. The ProVision CD included with this book contains:

- A demo version of ProVision;
- Several example inspection "projects";
- Images of objects to be inspected;
- An online manual;
- PowerPoint presentations of the VS710 Vision Sensor;
- Brochure on the VS710 system in PDF format.

To use the CD, insert it into the CD drive of your computer and execute the "start.exe" program. A graphical user interface will appear, enabling you to install the ProVision software or to view the presentations.

The user interface of ProVision is divided into a number of areas. As in most Windows applications, the upper section contains various toolbars and a menu and the lower area is a status bar, showing editing information. The main central part of the interface is for displaying the open "project". Open "projects" have left and right working areas. On the left side, the list of images to be processed and the inspection program are displayed in a tree structure. The right side displays the contents of the selected item from the left side. Processed and unprocessed images are also shown in the central part of the interface.

Inspection algorithms in ProVision are called *projects* and are saved in files with the VIS extension. A project consists of a number of *inspection elements*, which form an *inspection sequence*. Images to be processed are saved as bitmap files and so have the BMP file extension. Sample projects and images are located under the SAMPLES directory of the ProVision installation.

Inspection elements are divided into functional groups. A letter at the beginning of the element's name identifies its function group. The groups are:

- C – Circle functions for processing circular lines.
- H – Histogram functions for processing grey-level frequency distributions.
- I – Image capture elements for acquiring images.
- L – Line functions for processing straight lines.
- M – Miscellaneous elements, which include feature extraction and classification functions.
- P – Point functions for linking position points.
- V- Vector functions for processing straight lines.
- W – Window functions for processing rectangular image windows.
- C – Control elements for controlling program flow.
- I – Interface elements for outputting results via VS710 interfaces.

To run one of the sample projects from the CD, first start the ProVision software on your computer. Then, select PROJECT->OPEN from the pull-down menu. Choose one of the project files using the open file dialog. The program is run by selecting TEST->START from the menus. To view the values of the processing elements during execution, select the options BREAKPOINT AT END OF CYCLE and DISPLAY TEST RESULTS from the TEST menu. Double clicking on any inspection element will generate a dialog box, where the parameters of the element can be specified.

Creating a New Project in ProVision

A guide to creating a new project with ProVision can be found by selecting the GETTING STARTED item in the HELP menu of ProVision. In summary, the steps for this are:

1.   Create a new project by selecting the PROJECT->NEW menu item. A dialog box will then appear where the image to be processed in the project should be specified. There are two standard parts to a "project": **image acquisition** and the **inspection program**.

2.   Simulate "**image acquisition**" by listing the bitmap images to be processed.

3.   Generate the **inspection program** by adding inspection elements in sequence.   Inspection elements are inserted by selecting the IMAGE PROCESSING->INSPECTION ELEMENTS menu item and double clicking on the desired element.   Help for any element can be found by selecting it and pressing F1.   Configuration of the elements' parameters is done by double clicking on the element in the inspection sequence and entering the desired values into the pop-up dialog box.

4.   Execute the program by selecting the TEST->START menu item. Debugging and analysis are carried out using options from the TEST menu, such as BREAKPOINT, SINGLE STEP EXECUTION and DISPLAY TEST RESULTS.

Author Index

Index